Hot Science is a series exploring the cutting edge of science and technology. With topics from big data to rewilding, dark matter to gene editing, these are books for popular science readers who like to go that little bit deeper ...

AVAILABLE NOW AND COMING SOON:

Destination Mars:
The Story of Our Quest to Conquer the Red Planet

Big Data:
How the Information Revolution
is Transforming Our Lives

Gravitational Waves:
How Einstein's Spacetime Ripples Reveal the Secrets of the Universe

The Graphene Revolution:
The Weird Science of the Ultrathin

CERN and the Higgs Boson:
The Global Quest for the Building Blocks of Reality

Cosmic Impact:
Understanding the Threat to Earth from Asteroids and Comets

Artificial Intelligence:
Modern Magic or Dangerous Future?

Astrobiology:
The Search for Life Elsewhere in the Universe

Dark Matter & Dark Energy:
The Hidden 95% of the Universe

Outbreaks & Epidemics:
Battling Infection From Measles to Coronavirus

Rewilding:
The Radical New Science of Ecological Recovery

Hacking the Code of Life:
How Gene Editing Will Rewrite Our Futures

Origins of the Universe:
The Cosmic Microwave Background
and the Search for Quantum Gravity

Behavioural Economics:
Psychology, Neuroscience,
and the Human Side of Economics

Hot Science series editor: Brian Clegg

OUTBREAKS AND EPIDEMICS

OUTBREAKS AND EPIDEMICS

Battling infection from measles to coronavirus

MEERA SENTHILINGAM

Published in the UK and USA in 2020
by Icon Books Ltd, Omnibus Business Centre,
39–41 North Road, London N7 9DP
email: info@iconbooks.com
www.iconbooks.com

Sold in the UK, Europe and Asia
by Faber & Faber Ltd, Bloomsbury House,
74–77 Great Russell Street,
London WC1B 3DA or their agents

Distributed in the UK, Europe and Asia
by Grantham Book Services,
Trent Road, Grantham NG31 7XQ

Distributed in the USA
by Publishers Group West,
1700 Fourth Street, Berkeley, CA 94710

Distributed in Australia and New Zealand
by Allen & Unwin Pty Ltd,
PO Box 8500, 83 Alexander Street,
Crows Nest, NSW 2065

Distributed in South Africa
by Jonathan Ball, Office B4, The District,
41 Sir Lowry Road, Woodstock 7925

Distributed in India by Penguin Books India,
7th Floor, Infinity Tower – C, DLF Cyber City,
Gurgaon 122002, Haryana

Distributed in Canada by Publishers Group Canada,
76 Stafford Street, Unit 300
Toronto, Ontario M6J 2S1

ISBN: 978-178578-563-4

Typeset in Iowan by Marie Doherty

Printed and bound in Great Britain
by Clays Ltd, Elcograf S.p.A.

For my parents Amirtha and Sivam,
husband Ian, and son Reuban

ABOUT THE AUTHOR

Meera Senthilingam is a journalist, editor and public health researcher specializing in global health and infectious disease. She has worked with multiple media outlets including CNN and the BBC, and research institutions including the London School of Hygiene and Tropical Medicine and the Wellcome Trust.

CONTENTS

INTRODUCTION: ROOM 911

Every outbreak has its source, its origin, and its index case. It has a place and person where it all began and where history was made. In 2003, the outbreak was Severe Acute Respiratory Syndrome, SARS, which went on to become a global pandemic, a public health emergency of international concern – and it began with a doctor and a popular tourist hotel.

On 21 February, Dr Liu Jianlun checked in to room 911 of the Metropole hotel in the urban area of Kowloon in Hong Kong. Liu was in town for a family gathering, but was tired after an exhausting few months at the hospital where he worked in Guangzhou, China. A sudden outbreak of a pneumonia-like infection had struck the port city of Guangzhou, with the wider province of Guangdong seeing more than 1,500 cases since November the previous year. Liu himself hadn't been feeling well on his departure, but had persevered with his journey. Once in town, he chose to explore, as the city had changed since his last visit. But by the next day, 22 February, he was too ill to continue, experiencing a fever, shortness of breath and low oxygen levels. He admitted himself to the nearest hospital, Kwong Wah Hospital in Kowloon, where he would die nine days later.

For the virus that killed him, Liu had been the vessel as it embarked upon a global journey. In Hong Kong alone, 1,755 people were infected with the same virus and by July 2003, less than five months later, more than 8,000 people were infected across 32 countries and administrative regions worldwide, of whom more than 810 died.

Called to the scene in Kowloon to check on the 'very unusual patient' was Professor Yuen Kwok-yung, Chair of Infectious Diseases at Hong Kong University and a physician at Queen Mary's hospital on the Hong Kong Island side of the city. 'The patient was very sick and they were already putting up all the infection control measures,' said Yuen. Liu's brother-in-law, a Hong Kong resident, was soon admitted to the same hospital with similar symptoms, having come in contact with Liu (now known by all involved as the 'index case', the case that brought the infection to the attention of health authorities). Both patients were originally diagnosed with mild flu of unknown origin and given medication accordingly, but to no effect. This opened up a mystery for Yuen's team to solve.

In the hospital Liu had been working at in Guangzhou, more than 100 medical staff had become infected while treating patients, which Yuen describes as very surprising. 'Usually influenza is easily controllable, with masks for example,' said Yuen. 'But this was not.' A lung biopsy from Liu soon revealed something else was at play, a previously unseen disease. It would become known as Severe Acute Respiratory Syndrome (SARS), part of a family of viruses known as coronaviruses, which include the common cold. SARS had originated from an unknown animal, though bats are suspected to have transmitted the virus to the civet cat, which then spread it to humans. The challenge to control the virus began instantly, as Liu's night at the hotel had made this a global race. The virus was already in bodies and on planes headed to countries as far apart as Singapore and Canada, which would each see over 200 cases.

There was no vaccine or treatment for the disease. Instead, extensive global surveillance and coordination to quarantine cases and trace their contacts enabled the outbreak, now considered a pandemic, to end five months later.

But the memory of SARS was overshadowed at the end of 2019 by a viral relative that would wreak greater havoc across the planet. A new coronavirus emerged at a seafood market in the city of Wuhan, China, again from an unknown animal source, infecting over 9,800 people and killing over 210 in less than one month. The disease, named COVID-19, caused fever, tiredness and a dry cough and caused some people to have difficulty breathing, or worse, with the elderly being the most severely affected; it reached nineteen countries during that first month. The majority of cases, though – more than 9,700 – were in China, which experienced rapid spread and saw cases reported in every province within weeks, though one province, Hubei, remained the epicentre.

The city of Wuhan was put on lockdown, major airline operators cancelled flights into China, tourists in the country at the time were repatriated by their respective governments, borders were closed to Chinese nationals and airport checks were put in place across the world. Mobility was put on pause and the WHO warned of stigma emerging against the Chinese population.

But China and the world had learned from SARS: health systems had strengthened significantly and the Chinese government was ready to do whatever it took to stop another contagion, including building additional hospitals in a matter of days as the country's own health system became overwhelmed. By March, case numbers topped 100,000 (more than 80,000 of which were in China), with over 4,000 deaths, but this period was also a turning point in the outbreak, both within and outside of China. China saw the number of new cases reported each day beginning to decline, though controversy surrounded the transparency of officials and the changing definition of what constituted a case.

As China experienced a decline in new cases, however, new focal points emerged in other parts of the world, most notably in South Korea, Italy and Iran, which each saw several thousand cases by early March. Seventy-two countries were now reporting cases and a cruise ship named the *Diamond Princess* was forced to dock off the coast of Japan to quarantine its passengers, among whom 706 cases were reported, forcing countries to send repatriation flights to collect their citizens. The WHO instructed all countries to ensure adequate preparedness plans were in place, ensuring they were ready to manage imported and locally transmitted cases, with laboratories capable of confirming probable cases and hospitals prepared to isolate and treat patients accordingly.

More flights were cancelled and quarantine zones put in place; some borders were closed; health systems were restructured. Meanwhile, multiple research efforts were underway to develop a vaccine and therapeutics. Experts believed the decline in China should become possible elsewhere once a peak was reached and the virus could be contained.

Eventually. Meanwhile, the world awaited the arrival – and spread – of the virus in their region.

But one thing is certain: these coronavirus cousins have held the world to account. SARS set a global health precedent, teaching us that when it came to infections, there were no longer any borders or limits and the world needed to work together to fight, or ideally prevent, a worldwide pandemic. Almost twenty years later, COVID-19 showed we still weren't ready to do this as efficiently as we might, enabling a local outbreak to spiral into one significantly affecting the world at large.

TWENTY-FIRST-CENTURY INFECTIONS 1

“*Novel virus infects thousands in one day.*

Measles causes national emergency.

More Ebola cases reported in the Democratic Republic of Congo.

Seasonal 'flu predicted to cause thousands of deaths this winter.”

In this day and age, outbreaks are daily news across most regions, available to read about soon after they occur, reaching the mainstream when the cases, disease, or location are big enough. Today we face a plethora of potential infectious adversaries: new and ancient, unknown and reborn. But regularity and familiarity make an infection no less feared when it arrives on one's doorstep. Outbreaks continue to elicit panic in the majority. Viruses, bacteria, fungi, parasites, and other microorganisms harbour the ability to penetrate human barriers with ease and break through defences with the sole purpose of attack, even when they are being watched closely.

For centuries, humanity has fought contagion, working hard to catch, treat or prevent disease, but success has been limited, short-term, and any progress met with an onslaught of new arrivals – or the same enemy in new armour. The war continues today, as only one battle has truly been won to date, against smallpox, an ancient virus once feared throughout the world and thought to date back at least 3,000 years to the Egyptian era. The virus killed 300 million people in the twentieth century alone and its end is considered to be the biggest achievement in international public health. But most experts know another victory of this sort will be challenging, most likely impossible, as smallpox was a comparatively straightforward target. Multiple efforts beforehand had failed: first against hookworm, then yaws, then malaria, and current efforts to end two other diseases – polio and Guinea worm – are lagging decades behind.

The success of smallpox, however, set the world on a new trajectory where a disease could be destroyed, in this case using immunization as a valuable and strategic weapon. It's unlikely we will rid the world of all infections, but after the success of smallpox, there is at least motivation to try.

The beginning of the end

When one imagines an outbreak, one thinks of a disease surging through a population, knocking down everyone in its path with extreme, debilitating and often fatal symptoms. This was the reality of smallpox, known to some as 'the scourge of mankind', making it a priority to protect the world against it.

The virus behind the disease, variola, invaded almost anyone it came into contact with, causing a fever followed by a distinctive rash. Fluid-filled spots containing the virus would take over the body, bringing death to many and leaving survivors scarred. The disease killed approximately three out of every ten infected. The world was officially rid of smallpox in 1980 after an eradication effort that had begun thirteen years prior. But it could be said that the path towards ending the disease began almost 200 years earlier, on the arm of an eight-year-old boy named James Phipps.

Phipps was the son of a gardener who just happened to work for Edward Jenner, an English physician and scientist who would come to be known as the father of immunology. For years, Jenner had heard rumours that milkmaids exposed to cowpox become naturally protected against smallpox. The cowpox virus closely resembles variola and word had spread that humans who came into contact with cowpox developed a milder disease they soon recovered from, which left them immune to its more fatal relative.

With experimentation being more rogue and unrestricted at the time, Jenner decided to test the theory that cowpox could be given deliberately to humans as a means of protection – and Phipps would be his proof. In May 1796, Jenner found milkmaid Sarah Nelmes, a recent cowpox patient. Nelmes apparently caught the virus from a cow called Blossom (whose horn is now on display in Jenner's house in Berkeley, Gloucestershire; her hide is kept in the St George's Medical School library in Tooting, south London). Jenner sampled the virus lurking in Nelmes's lesions and used it to inoculate young Phipps. The eight-year-old went on to develop a mild fever and loss of appetite but recovered after ten days. Two months later, in July, Jenner exposed Phipps to smallpox and no symptoms or lesions developed. He appeared to be protected. Jenner went on to successfully repeat this on more people over the following two years, again poor labourers, their children or inmates of workhouses. In the coming years, however, people across all classes were inoculated and vaccination (after the Latin for 'cow') soon became widely accepted.

The idea that our bodies could be protected against infection, using an infection, was born and smallpox would eventually become the target of a global defence, creating a biological shield that would one day span across the planet. But the road to get there would be far from straightforward.

Ending smallpox

An intensified programme to eradicate smallpox began in 1967, when there were still more than 10 million cases occurring across 43 countries. By this point, the disease had already been eliminated – meaning it had stopped spreading in a particular geographical region – in North America and Europe, following an initial effort to end the disease, launched by the World Health Organization (WHO), in 1959. The programme had focused on vaccinating the masses and the target had been to get at least 80 per cent vaccine coverage in every country in order to reach the herd immunity threshold, a level of coverage where the chances of unvaccinated people getting the disease is extremely low. (The threshold varies from one disease to another, based on how easily the infection transmits between people.) But South America, Asia and Africa continued to see millions of cases, while Europe and North America were still seeing imported cases, particularly as air travel rose in popularity. Governments across all regions were therefore motivated to end the disease, which would require a change in strategy.

'Little attention was paid to the reporting and control of cases and outbreaks, which we felt were the most important things,' the late Dr Donald A. Henderson told the WHO in a 2008 interview. Henderson, who died in 2016 at the age of 87, had led the international effort to end smallpox despite criticism that it was an impossible task. He had emphasized that simply vaccinating everyone against smallpox, or any disease, was not always feasible and therefore should not be the sole strategy. Public health teams needed to understand the severity of the situation – how many were affected and where – to better target their resources as well as to contain those infected to stop them spreading the disease further. 'We made a very strong point about the need for surveillance of cases and their containment,' he said.

Epidemiologist Dr William Foege soon implemented a surveillance and containment strategy under Henderson's leadership and significant reductions were quickly accomplished. For example, Foege's team used limited resources to focus solely on outbreak-affected areas when working in eastern Nigeria in 1967, identifying cases and vaccinating everyone within a defined radius of an outbreak, known as ring vaccination. This curtailed the outbreak within five months, despite just 750,000 people of the 12 million people living in the region receiving the vaccine.

The method was proven to work again and again and was simply much more efficient than trying to reach everyone, says Dr Donald Hopkins, who led the effort to control smallpox in Sierra Leone in 1967. The disease typically affected 5 per cent of a population at most at any one time, Hopkins explains, so the aim was to identify that 5 per cent and focus efforts there. Using this approach, Hopkins' team saw results within months in Sierra Leone, and within a few years in the

West African region as a whole, despite the region having the worst infrastructure of any they were working in globally.

The situation was much more difficult in populous India, Hopkins notes, as four years after success in West Africa, teams were still failing see an impact there. Government teams set out to visit every household in the country in the space of ten days in 1973 to identify the true extent of cases and stop the disease spreading more quickly. Some states were found to have twenty times more cases than previously reported. Once this was known, resources were deployed with greater accuracy and India reported its last case of smallpox about one year later. This surveillance-based approach has since become the backbone of outbreak control.

'The concept was that if we could discover the cases more quickly than before, the containment teams could interrupt the chains of transmission,' Henderson told the WHO, explaining teams could then break those chains by vaccinating possible contacts in areas where there were cases. Additional factors further aided the success of the global eradication campaign: for example, community teams ventured out to people rather than waiting for them to come to health facilities, meaning they reached everyone, including the most remote. A further development was the introduction of a bifurcated needle in 1968, a thin metal rod with two prongs that would hold a dose of the vaccine for more efficient injection into the skin. The beauty of this ingenious tool was its simplicity when compared to the jet injectors previously used; it enabled 100 doses to be delivered from a single vial.

But despite these factors coming together to bring immediate progress in reducing cases, the road to eradication took twelve years, with societal and political elements coming into play, such as civil wars, extreme weather, poor infrastructure and, importantly, convincing people the vaccine was safe. As these hurdles were overcome and cases of smallpox decreased, the need for resources in order to find the remaining cases increased. The cost per case rose significantly as teams travelled farther and wider to find the final few. But they did find them.

On 26 October 1977, experts found and isolated the last case of naturally occurring smallpox in 23-year-old hospital cook Ali Maow Maalin. Maalin, from Merca, Somalia, had worked as a vaccinator in the smallpox programme, yet had avoided the vaccine himself due to a fear of needles. The virus caught up with him as he helped direct a driver taking two children to a nearby smallpox isolation camp. Nine days later Maalin developed symptoms. He was told to stay home while special teams were sent to vaccinate the households around his location, successfully reaching more than 54,000 people in two weeks. Maalin recovered and smallpox was over. Almost.

Janet Parker, a medical photographer, and her mother were the last official cases of smallpox, in August 1978. Parker is presumed to have contracted the virus at Birmingham University medical school, where research on the virus was taking place. She became ill on 11 August and developed the signature rash on the 15th, but was not diagnosed with smallpox until nine days later. She died on 11 September. Her mother, who had been caring for her, also contracted the disease, but survived. Parker had access to the laboratory where a smallpox specimen was contained, but it is unclear if she was infected by entering that laboratory or by the ventilation system taking the virus from the laboratory to her office in the same block, explains David Heymann, Professor of Infectious Disease Epidemiology at the London

School of Hygiene and Tropical Medicine, who worked with the eradication programme for two years.

This event instigated a programme of consolidation or destruction of remaining specimens of the virus, explains Heymann. This took place during the Cold War, and countries were offered the choice of giving their smallpox stocks to the United States, to the USSR, or destroying their stocks under a set protocol. Both the US and Russia continue to hold on to the stocks given to them, monitored and handled by the World Health Organization under a WHO agreement set in 1979. The US stock is held at the Centers for Disease Control and Prevention (CDC) in Atlanta and the Russian stock at a research laboratory in Siberia. The two facilities are inspected by the WHO every two years.

Debates on whether these last remaining stocks of smallpox should be destroyed have been ongoing for decades. Research on the virus continues today, following anthrax attacks in the United States in 2001, during which anthrax spores were found lacing mail sent to news agencies and congressional offices. This led to the development of bioterrorism preparedness programmes, which include research on smallpox, looking for better diagnostic tests, antiviral medications and safer vaccines. But gene technology has enabled the smallpox virus to be fully sequenced, meaning that the virus can be reconstructed for research purposes if needed, which some experts, including Heymann, believe removes the need for live virus stocks to be stored. The World Health Assembly, the decision-making body of the WHO, has requested the review of research using the live smallpox virus on multiple occasions, with the 69th assembly calling upon an Advisory Committee to review this in May 2016. At the 72nd assembly in 2019, the committee stated that research using the virus was still needed to continue the development of antiviral medicines for smallpox preparedness, so the stocks remain.

Officially, though, smallpox has been eradicated, with Ali Maow Maalin being the last face of the smallpox pandemic that plagued the world for millennia. The eradication laid the groundwork for the routine vaccination programmes now implemented globally, in the WHO's Expanded Programme on Immunization, protecting children from multiple childhood diseases including measles, polio and tetanus. Could these too be eradicated? Most experts say no, but polio efforts are underway. 'Even towards the end of smallpox eradication, the senior staff never talked about potential eradication of any other disease,' Henderson told the WHO.

Smallpox had many factors on its side: the vaccine was heat stable and did not require refrigeration for storage, an invaluable property in the remote, tropical settings in which the vaccine was used; immunization required just a single vaccine dose; and everyone who had the disease could be identified by its distinctive rash and therefore easily isolated and their contacts vaccinated. Other diseases do not have this combination of winning elements; some have just a few, and most only one or two. But as pandemics and global health emergencies increase in frequency, how can teams not hope to at least try to rid the world of some of them for good?

When to reach for zero

Eradicating a disease means permanently removing all traces of its pathogen, be it a virus, bacteria, parasite or any other infectious microorganism. It involves bringing the number of pathogens found anywhere on the planet down to zero, then keeping them at bay, forever. As desirable as it may be to achieve this for every disease, the reality is that precise scientific and political criteria are required to even attempt it and very few infections fit the bill. At the moment, the International Task Force for Disease Eradication has identified eight possibilities: Guinea worm, also known as dracunculiasis, poliomyelitis (polio), mumps, rubella, lymphatic filariasis, cysticercosis, measles, and yaws.

The scientific criteria include a disease being epidemiologically vulnerable to attack, such as it not spreading easily, being easily diagnosed, or an infection leading to lifelong immunity in those who survive. An effective intervention must also be available, such as a vaccine or a cure to halt the disease in its tracks. Some evidence must also be available of the disease having been eliminated already in a particular region, demonstrating the possibility of its removal on a wider scale. With the exception of not spreading easily, all of these points applied to smallpox.

Politically speaking, governments worldwide need to care about the disease, its impact, and possible harm; eradication efforts need to be affordable and cost-effective; and complete removal of the disease needs to have a significant benefit over simply controlling the disease long term. 'The motivations of countries play into it,' explains Hopkins. In an ideal scenario, eradication efforts would also fit into other health programmes, or vice versa, to provide maximum healthcare benefits to people when they are reached. With all that in mind, experts within the task force identified the above eight diseases that show promise for eradication, but today just two have official programmes in place: Guinea worm and polio, since 1980 and 1988, respectively.

Yaws, a chronic, debilitating, bacterial infection that affects the skin, bone and cartilage, was among the first diseases targeted for eradication, in the 1950s, as it can be cured with inexpensive antibiotics. But success was limited. 'That programme really never finished,' says David Heymann of the London School of Hygiene and Tropical Medicine. 'I think it just stopped because of lack of engagement of the global community in eradicating it,' he adds, highlighting that cases were dotted around the globe and typically infected neglected populations, such as the Pygmies of Central Africa. Efforts to eradicate the disease were renewed by the WHO in 2012, but yaws is still prevalent, with more than 80,000 suspected cases in 2018, though just 888 were confirmed using laboratory tests. Fifteen countries remain endemic for the disease, meaning they experience continuous transmission.

Guinea worm is possibly one of the most gruesome diseases you could imagine, where drinking contaminated water leads to the development of a metre-long worm, or many worms, inside the body, which then burst out of the skin one year after infection. More on this disease in chapter 3, but its easy diagnosis, possibilities for prevention through simple interventions to stop people drinking contaminated water, and the economic benefits to be gained from its demise made it a top candidate for eradication soon after the victory over smallpox. The recent discovery of the parasite also infecting dogs, however, has recently brought the idea of eradication into question.

Polio then showed promise as it had an effective vaccine, immunity against the disease was lifelong, and the disease only affects humans, meaning there were no animals to deal with that could act as reservoirs or spread the disease, as is the case with malaria, for example. Elimination of polio had also been achieved in the Americas, showing it was possible to stop transmission of the disease. But complete eradication of polio would be far from easy, as the vaccine is not heat stable and requires multiple doses. (More on the successes and struggles to date in chapter 3.)

While these two programmes continue, the World Health Organization also has a number of others underway working to eliminate, not eradicate, diseases such as rabies, leprosy (yes, it still exists), and the blinding eye infection trachoma. 'In general, elimination doesn't mean getting to zero, so it's by definition a much easier target than zero,' says Hopkins. But it depends on how you define elimination, he points out, be it stopping a disease in an area or reducing levels of the disease to a manageable level. Hopkins believes the important thing is to reduce the greatest amount of suffering. In some instances, this only requires control of a disease, in others its elimination and in a few instances its eradication, which will be difficult and expensive. Each disease is unique and for humankind it's about knowing your enemy, picking your battles and knowing when to sign a treaty, fight regionally or declare a world war.

This is an emergency

A century ago, from 1918 to 1920, the world grappled with the most severe and deadly pandemic in recent history. An H1N1 influenza virus, thought to have originated from birds, swept across the planet infecting 500 million people – one third of the world's population at the time – and killing 50 million people, according to the US Centers for Disease

Control and Prevention. This was the Spanish flu pandemic, in which infections were powerful enough to kill young, healthy adults, not just the elderly and infirm as per regular, seasonal influenza. Striking at a time when vaccines and treatments were not available and a world war was underway, the virus unsurprisingly had devastating consequences as the only weapons to hand were isolation and quarantine along with attempts to promote good hygiene. H1N1 would leave its mark as something to be feared throughout history, ending in 1920 with no clear understanding about how it was stopped.

In 1957, another influenza strain emerged in East Asia, H2N2, triggering a pandemic known as the Asian flu, first reported in Singapore and spreading as far as the United States, killing an estimated 1.1 million people worldwide. Yet another strain, H3N2, began spreading in the United States in 1968, killing a further 1 million people worldwide. These latter two pandemics are lesser known, but their damage was extensive and reminded populations at the time of the damage a new infection, particularly influenza, can do.

In 2003 we saw SARS, a previously unknown virus, surge across an unprotected population to infect thousands across Asia and reach all corners of the globe. Six years later, in 2009, we were reminded of the power of influenza as a new form of H1N1 emerged in Mexico, soon reaching the United States, Canada and promptly the rest of the world. The severity of the pandemic was lower than predicted, but 214 countries globally reported cases and at least 18,500 people died, though some studies show this to be a vast underestimate, with one study by the CDC suggesting fifteen times more people died, an estimated

284,000. This time the world had antiviral treatments and vaccine technologies as well as international agreements to help countries work together to curtail the spread, but there was a new societal norm to contend with: population mobility. People harbouring the virus were on planes and trains travelling to new destinations before they even knew they were sick, helping H1N1 reach six continents within just nine weeks of it first being reported. (More on this pandemic in chapter 7.)

Most recently, as 2019 gave way to 2020, a novel coronavirus emerged in the populous city of Wuhan, China, infecting more than 100,000 people and killing over 3,500 by early March. Though initially the majority of cases occurred within the country, with transmission aided by its timing during the lunar new year, international spread of the virus promptly became a concern as over 20,000 cases were reported outside of China: across more than 100 countries by this point. Advances in technology meant the virus was identified within weeks and multiple efforts to produce a vaccine, using the DNA of the virus, were soon underway. However, China had to battle its worst outbreak ever and soon after so did many other countries, notably Italy, South Korea and Iran. As a result, air travel was restricted and people the world over were advised not to travel to these countries, while those already there were told by their respective governments and health authorities to leave and self-isolate. Cases from China created hotbeds in other countries, from which people soon travelled to tens of other countries worldwide.

According to the Swedish flight information service Flightradar24, more than 200,000 flights take place globally every day, and the World Tourism Organization recorded over 1.3 million international tourists arriving at borders in 2017.

The world is increasingly becoming one big canvas that is easier to explore with the regular arrival of new people in new places for the long and short term. With this comes exposure to new, previously unknown microorganisms that wreak havoc among an unsuspecting, and unprotected, population. The speed at which aircraft transport people from A to B, even from one end of the planet to the other, is shorter than the incubation period for many infectious diseases, meaning someone can become infected and be in a new location, among vulnerable people, before showing symptoms and transmitting the infection.

Air travel was a key player in the 2003 SARS pandemic, for example, resulting in the virus reaching a range of new countries within days of emerging. The pandemic led to the revision of the International Health Regulations (IHR) in 2005, an international agreement among 196 WHO member states, first implemented in 1969 when the scope was limited to just four infectious diseases: cholera, plague, yellow fever and smallpox. The 2005 updates aimed to address modern-day needs, with the goal of helping to 'prevent and respond to acute public health risks that have the potential to cross borders and threaten people worldwide,' according to the World Health Organization. The regulations include provisions for outbreaks that are a 'public health emergency of international concern' (PHEIC), based on an unexpected or sudden situation that could cross borders and therefore become an international concern, requiring a coordinated international response. A PHEIC is decided on by an emergency committee made up of relevant international experts and a declaration means countries must be ready to work together to stop the outbreak by providing vaccines, teams on the ground, funding, or resources to prevent border closures, avoiding the detrimental effect this can have on countries' economies.

Since SARS surprised and scared the planet, six emergencies have been declared to be of international concern. The first was the 2009 H1N1 swine flu pandemic, followed by polio in 2014, the 2015 Zika virus epidemic, two outbreaks of Ebola – one that hit West Africa in 2014 and another in the eastern region of the Democratic Republic of Congo that began in 2018 – and the novel coronavirus outbreak that began in China at the end of 2019. Four very different families of viruses, given SARS was also a coronavirus, but all deemed to be a significant risk to the global population if not managed appropriately. Air travel and transport has been a key factor in their spread, but just as influential have been rising global temperatures, the increasing mobility of populations, and a rise in urban or peri-urban populations living close together, often with poor infrastructure and poorly managed health services around them. Dense, poorly planned urban environments provide the perfect ecosystem for infections to be introduced and then thrive, explains Dr Mike Ryan, Executive Director of the Health Emergencies Programme at the WHO.

Ryan points out that there are plenty of public health emergencies to worry about, not just those declared to be an international concern. His department is informed of around 7,000 public health events each month, the majority of which are infectious outbreaks, he says. In 2018, 481 events were considered significant enough to warrant help from the WHO, be that minor administrative support or a full-scale response. At any one time, his teams are handling tens of emergencies, two-thirds of which take place in fragile states within the Middle East and Africa, with the most common outbreaks being cholera, measles, diphtheria and yellow fever. Ryan mentions measles and diphtheria with dismay as both have a cheap and effective vaccine, meaning they should not be causing outbreaks, but weak health systems and poorly structured national health programmes along with increasing hesitancy towards vaccines are seeing these preventable infections return.

Further outbreaks of concern involve animals that spread the disease and flourish in dense populations, including mosquito-borne diseases such as dengue fever, yellow fever and chikungunya. In some cases, such as Ebola, the outbreaks originate from wild animals living near rural populations but are then 'driven by the health system itself', says Ryan, given the lack of basic infection control being practised and unreliable resources.

Uncommon, emerging, diseases are a further worry among public health officials as the element of surprise they bring means countries will lack effective drugs or vaccines and be unprepared if they strike. SARS and the novel coronavirus COVID-19 are prime examples of this. In an attempt to be ready for future unknowns, the WHO maintains a list of emerging diseases it considers to be a priority for research and development due to their 'potential to generate a public health emergency, and for which insufficient or no preventive and curative solutions exist'. Currently, these are Crimean–Congo Haemorrhagic Fever (CCHF), Ebola Viral Disease and Marburg Viral Disease, Lassa Fever, Middle East respiratory syndrome coronavirus (MERS-CoV) and Severe Acute Respiratory Syndrome (SARS), Nipah and henipaviral diseases, Rift Valley Fever (RVF), Zika disease, and 'Disease X' – meaning a completely new, previously unseen infectious disease, such as COVID-19 at the time of its emergence.

The chosen diseases range significantly in their effect, transmission and severity. Diseases like Ebola spread readily from person to person when in close contact and strike quickly and fatally, while Zika is transmitted by mosquitoes and causes long-term disease more than it does fatalities. SARS spread quickly and was fairly deadly, with a fatality rate of approximately 10 per cent, while MERS does not spread as easily but is more fatal to those it strikes. COVID-19 appears to spread more easily but is less fatal. It's also worth noting the list is composed entirely of viral infections, which some experts believe comes down to viruses typically being smaller in size and more likely to spread via droplets in coughs and sneezes, for example, helping them spread much more easily than other microorganisms.

MERS has an unusual outbreak history. An acronym for Middle East Respiratory Syndrome, the coronavirus behind the disease is in the same family as SARS and COVID-19. As its name suggests, MERS is most common in the Middle East where it was first identified in 2012, but the first outbreak of global concern was in 2015 when a Korean man contracted the virus while visiting the region on a business trip. He returned home to South Korea, sparking an outbreak that would go on to infect 186 people, killing 36 of them. The outbreak came as a surprise to global health officials as South Korea has a good health system, explains Dr Sylvie Briand, Director of the Infectious Hazard Management team at the WHO. It was a shock for both the country's health sector and global agencies.

The problem here was panic after the continent's experience with SARS a decade earlier, along with a slow pace at identifying and isolating cases and tracing contacts, but also a cultural pattern of 'doctor shopping', says Briand, as people tend to visit multiple doctors for second and third opinions and in doing so spread their infection across healthcare facilities. In addition to traditional surveillance and prevention, her team advised the South Korean government to ask people to stop visiting multiple doctors. They also set up a Korean Centers for Disease Control, a dedicated agency for outbreaks, which combined with a more prepared health sector meant an imported case in 2019 was promptly controlled. The 2015 outbreak cost South Korea's economy $2 billion as travel to the country declined, people stopped going out to restaurants and indulging in other leisure activities and schools closed, with parents then keeping children home even once they reopened.

When asked if outbreaks, and in turn global pandemics, are becoming more common, both Briand and Ryan believe the answer is yes. 'Society is evolving so fast and we need to keep adapting,' says Briand, adding that the extent to which a disease spreads 'really depends on the pathogen'. Some are highly contagious and others require vectors: animals that transmit the disease from person to person, such as mosquitoes. In such cases, the impact made by the disease depends on how the vector thrives. But Briand is hopeful, believing that new technologies will help to contain them earlier. 'It's a matter of balance,' she says 'between the natural phenomenon and measures we can put in place to mitigate these phenomenon [*sic*].' She adds that society will keep moving forward and cannot go back, or be controlled, for example regarding urbanization. Big cities are hubs for mobility, with train stations, airports, bus stations, with a lot of in and out movement, meaning 'preparedness is important'.

Data as of May 2019 reported to the global pandemic monitoring board showed that more than 100 countries worldwide had low or moderate levels of national preparedness against pandemics. Just nineteen countries had reached the top level of preparedness, meaning a national capacity greater than 90 per cent in handling factors ranging from zoonotic (animal) diseases and antimicrobial resistance, to legislation, national laboratory networks, emergency response operations and risk communication, to name a few. The list of nineteen included Australia, Canada, China, South Korea, the United States of America, Cuba, Finland, and the UK.

As countries become stronger and more prepared, outbreaks are not increasing in number at the rate they have been in recent years, explains Ryan, but overall, outbreaks are nonetheless continuing to rise. 'The pattern is clear,' says Ryan, in that fragile states, failing health systems, vaccine hesitancy, pathogens resistant to the drugs used to treat them, mobile populations and climate change are setting new patterns that 'we're not catching up with fast enough'. He believes the world is waking up to the reality of climate change, but not yet the ecosystem resulting from that; Ryan explains that 'we have created the perfect conditions for an emerging disease to spread across the world.'

His team will continue to deal with public health emergencies to mitigate their impact, using the technologies available to them, he says, but he noted that 'somewhere around that corner that we can't see is the next pandemic, the next SARS' – and now his team are leading the global fight against COVID-19.

Asked if the world was ready for it, Ryan didn't believe it was. 'We could be a lot more ready if we really wanted to be,' he said, stating that we just hadn't taken the threat of that seriously enough. Until now, that is.

Outbreak? Epidemic? Pandemic?

We're dealing with an epidemic, news reports say, with respect to diseases such as malaria, Zika, a regional outbreak of cholera, but also obesity, diabetes, mental health and habits such as vaping, opioid use, and now screen time among children. What began as one case, or a cluster, grew to become an outbreak, and soon an epidemic. If not controlled it could go on to become a pandemic, or over time, endemic. But what does this all really mean? Each word, each stage, has a specific and structural meaning with origins in infectious disease control but as a society we have grown to use many disease control terms in broader conversation and as a result, often employ them incorrectly, inaccurately, or with flippancy.

It all starts with **health**, defined by the WHO as a state of complete physical, social and mental well-being, and not merely the absence of disease or infirmity. Then comes **public health**, a seemingly straightforward term, but not quite. More accurately it is a social and political concept, WHO glossaries state, aimed at improving health, prolonging life and improving the quality of life among whole populations using interventions such as disease prevention, healthcare policies or health promotion programmes that are, importantly, organized through society.

Every disease fits into one of two camps: **communicable** and **non-communicable** disease, with the term communicable meaning the disease can be spread from one person or animal to another. Communicable diseases include the common cold, influenza, HIV, malaria, tuberculosis, plague, to name a few. Non-communicable diseases include cancer, diabetes, asthma and cardiovascular conditions. A communicable disease, also known as **infectious**, can spread between humans either directly, through coughs, sneezes and personal contact or indirectly, through animals, such as mosquitoes, or contaminated food or water. Some diseases spread directly from animals to humans and are known as **zoonotic diseases**, or **zoonoses**. The category typically only includes diseases spread from vertebrate animals: Ebola, for example, is introduced into human populations by contact with infected wild animals such as monkeys or fruit bats, and influenza viruses are known to cross over from animals, including birds. All communicable diseases are infectious, but only diseases that spread by direct contact from person to person can be called **contagious**, a common mistake people make, using the two terms interchangeably. All communicable diseases are spread through a pathogen, a bacterium, virus, parasite or any other microorganism capable of spreading disease.

When working within public health to control either class of disease, teams need to understand the **epidemiology** of the disease at hand, which means understanding the patterns of a disease among populations to know how it affects different groups of people and why. Knowing this is the first step in planning how best to stop it. For example, malaria control strategies vary by geographical region as different parasites and mosquitoes are involved in its spread in different locations. Malaria is caused by one of four types of *Plasmodium* parasite: the parasite is spread among humans through the bite of infected female *Anopheles* mosquitoes, different species of which can be found on different continents. The region most affected by malaria is sub-Saharan Africa, where the *Plasmodium falciparum* parasite dominates and is spread through *Anopheles gambiae* mosquitoes, which bite primarily at night, meaning the provision of insecticide-treated bed nets is an effective strategy in protecting people there.

A person infected with a disease is known as a **case**. When an outbreak occurs, it begins with the first case noticed by health officials, the person who first indicates the presence of an outbreak, known as the **index case**. This is not to be confused with the **primary case**, which is the first person to bring an infection into a group of people or community, but may not be the first person brought to the attention of health officials to signify an outbreak has begun. Which brings us to the most popular word in the field of disease control: **outbreak**. Officially defined as the 'sudden appearance of a disease in a specific geographic area or population' by the US Centers for Disease Control, others including WHO expand on this to define it as cases of a disease being 'in excess of what would normally be expected in a defined community, geographical area or season'. For example, a town may see a few cases of measles every now and then, but a sudden group of cases being reported would be an outbreak. With rare diseases, just two linked cases could be sufficient to be called an outbreak.

The next step up from an outbreak would be an **epidemic**, which carries the same definition regarding there being more cases of a disease than would be expected, but typically applies across a larger geographical area and often involves the rapid spread of the disease or issue across a population. The millions of people living with HIV globally constitute an epidemic, for example, as do the millions that continue to become infected with malaria and tuberculosis each year. A sudden rise in cases of any of these diseases in a particular location over a period of time would also be called an epidemic.

Some countries and regions are also **endemic** for certain infectious diseases, meaning they have continuous transmission of the disease. Malaria is endemic in more than 80 countries around the world according to the WHO, mostly in Africa, Asia and the Americas, while the eye infection trachoma is endemic in more than 40 countries, causing blindness or visual impairment in almost 2 million people. When endemic countries experience sudden outbreaks or epidemics, control efforts become more complicated, and political, with routine control measures needing to be supplemented with more rapid response strategies. Here, simply identifying an outbreak can be complex as it could be easily missed or disregarded as a new, or separate, concern.

At times, the term **cluster** may also be used, referring again to the number of cases of a disease being above what is 'normally expected' but with cases being aggregated, close together, in time and location. Once investigated, they may go on to be classed as outbreaks if considered to be a cause for concern.

The word that invokes fear in almost everyone is the big one: **pandemic**. Defined by the US CDC as 'an epidemic occurring over a very large geographic area', the WHO adds the detail of it being the worldwide spread of a 'new disease'. This latter point would explain why diseases with global epidemics, such as malaria, tuberculosis, and dengue fever, for example, are not considered pandemics. SARS and H1N1 swine flu were pandemics, as was the 1918 Spanish flu, also an H1N1 influenza virus. Most recently, the WHO added COVID-19 to the list. The finer points of what constitutes a pandemic are, unsurprisingly, unclear and are debated by scientists and health agencies alike, with experts arguing that factors such as novelty, severity, how easily a disease spreads, and attack rates – the proportion of people who fall ill once exposed – should come into play.

But the core of a public health response is about control: stopping transmission of a disease to prevent further spread, and to do this requires understanding of the bug behind it all – inside and out – and its timings. The time between someone being exposed to an infection and becoming infected is known as the **latent period**. The bug has not yet caused symptoms and typically cannot yet be transmitted to others. The time between exposure and developing symptoms of the disease is the **incubation period**. These first two periods overlap and are important to understand when it comes to disease control as cases may arise in new places, among newly susceptible people, before they realize they are infected. Importantly, when monitoring those at risk and putting people under quarantine, the incubation period determines the amount of time they are monitored for to determine if they have been infected or not. Measles, for example, has an incubation period of ten days, Ebola up to 21 days, and tuberculosis up to ten weeks.

The period when an infected person can transmit the disease is known as the **infectious or contagious period**. This can be before or after symptoms begin, depending on the disease. Many diseases are transmitted only once symptoms begin, so the contagious period begins after the incubation period – examples include Ebola and tuberculosis. Measles can be transmitted once first symptoms begin, such as a fever, but four days before a rash develops – and for four days after the rash has gone. Bacterial meningitis can spread seven days before symptoms begin, meaning the latent and incubation periods overlap. Understanding this for a target disease informs contact tracing in order to identify who else may be at risk.

Think of it all as a big puzzle with pieces of varying shapes and sizes. First the pieces need to be found, like identifying the pathogen responsible and figuring out how and when it transmits – and how viciously – then those pieces need to be promptly put together to form the bigger picture, allowing you to understand what you're dealing with and mount a response.

Once a disease is established and becomes an epidemic, two key terms help to explain its scale and continued growth, particularly for longer-running epidemics: the **prevalence** and **incidence**. The prevalence is the total number of new and existing disease cases within a population over a given time period, while the incidence is the number of new disease cases reported over a period of time. Both terms help to quantify a disease. In 2018, the global prevalence of HIV was 37.9 million, meaning this many people were living with the infection. Within that number, the number of new infections, the incidence, was 1.7 million in 2018, itself a staggering number. When handling outbreaks, the time period that the disease incidence is calculated for is smaller – often days, weeks or months – to identify how the outbreak is spreading and expanding and the level of resources needed to control it, such as vaccine or treatment supplies. For example, during the 2019 Ebola outbreak in the Democratic Republic of Congo, declared to be a public health emergency of international concern, a weekly situation report issued on 19 September reported 57 new confirmed cases during the week prior, as well as 145 cases, broken down by health zone, in the three weeks prior and a total of 3,034 confirmed cases since the outbreak began in 2018.

When controlling an infectious disease, the first steps involve isolation and quarantine of cases to limit the infection's ability to spread, contact tracing of people who may have been exposed to the infection through contact with known cases, and basic hygiene measures and treatments, if available, to further curtail the spread. But an ideal control strategy would also include the availability and use of a vaccine against the disease. Some diseases have fully effective vaccines available, such as polio, measles and mumps, and others have partially effective vaccines available, such as seasonal influenza, for which seasonal vaccines reduce the risk of contracting the flu by up to 60 per cent. But the majority of diseases, particularly new and emerging diseases, either do not have a vaccine or have some options still at the research stage, not advanced enough for use.

For diseases that are spread from person to person and have effective vaccines to hand, a key part of control efforts at the population level involves mass vaccination in order to reach the **herd immunity threshold**. This is a level at which a disease pathogen can no longer spread throughout a population – a herd of people – because enough of them are protected by the vaccine to block the pathogen transmitting to the number of people it needs to reach in order to survive. Its chain of transmission is blocked. The threshold varies for different diseases as it depends on how contagious the disease is. For example, measles is extremely contagious, as one infected person could transmit their infection, on average, to between twelve and eighteen unprotected people, while polio is less infectious, with an infected person able to spread the virus, on average, to between five and seven unprotected people. Measles therefore requires greater coverage of a vaccine throughout a population or community to stop transmission, resulting in a herd immunity threshold of approximately 95 per cent. If 95 per cent of the population has been vaccinated; the virus should not be able to spread. For polio, the threshold is around 80 to 85 per cent. The idea behind the strategy is that by getting vaccinated, people are protecting both themselves and the people around them. As we have seen, this was crucial in the eradication of smallpox, where at least eight in every ten people needed to be vaccinated to stop transmission.

How infections are borne

It could be argued that the most important factor we need to understand about a disease in order to control it, and end it, is how it spreads: its transmission route. This determines how and where an outbreak occurs and how it then rampages through a community as it grows. Its journey is by no means straightforward; with some diseases having multiple routes of transmission to ensure the infection spreads, and therefore survives. After all, that's what an infection is all about: keeping the pathogen going, entering new hosts, and ensuring its legacy. Complicating things further are transmission cycles, which occur when an intermediary is involved in spreading an infection, such as a mosquito. Add to that the fact that some diseases infect animals as well as humans, giving the potential for reservoirs of the infection to remain once all human cases are treated, or have died. But first things first: the many ways infections are borne.

There are six main ways in which infectious diseases transmit. These are through direct contact with a person's bodily fluids; through the air; via water; via oral contact with contaminated faeces, known as faecal–oral; via a vector; and from mother to child, either in the womb, during childbirth or through breastfeeding. The infections that spread through the air, such as measles and influenza, are typically the most infectious, though blood-borne infections, such as HIV, can be directly transferred from one person's blood to another, meaning that though the likeliness of this happening is smaller than airborne spread, the chances of transmission are higher when it does occur. The most complex transmission, perhaps, is vector-borne, as it involves a pathogen having life cycles within both humans and the vector in order to transfer from one to the other.

Direct person-to-person contact is how many may picture the notion of contagion; an ill-timed touch with someone harbouring a potentially fatal disease. Transmission occurs through bodily fluids such as blood, open sores, saliva or fluids involved in sex, like semen and vaginal fluids. Microorganisms that spread in this way are usually unable to survive for long once outside of a host body, hence the need to transmit directly. Diseases that fit into this group include sexually transmitted diseases such as gonorrhoea and syphilis, Ebola (through all bodily fluids), chickenpox (through open sores), herpes (through open sores) and conjunctivitis (through discharge from the eye). Some diseases have multiple routes of direct transmission, such as HIV, which can spread through blood and semen, vaginal fluids and breast milk.

Many diseases are transmitted through the air, including influenza, tuberculosis, whooping cough (also known as pertussis), and the common cold. Transmission between two people can happen in one of two ways, depending on the disease: an infected person coughs up or sneezes droplets that carry the pathogens short distances to then land on the eyes, ears or nose of a susceptible person (this happens, for example, with meningitis); or the infected person creates a spray that contains the pathogen as they cough, sneeze, talk or breathe (this happens, for example, with measles and tuberculosis), enabling the germs to spread over greater distances and to be inhaled by a susceptible person. The measles virus can persist in the air for up to two hours.

The containment of airborne diseases and many, but not all, diseases transmitted by direct contact requires the immediate quarantine of cases and tracing of people who have been in contact with cases, as well as good hygiene practices and, in the case of airborne diseases, ventilation. Infections such as HIV or other sexually transmitted diseases do not require quarantine as transmission is dependent on the infected person having sex, which they will be advised against doing before they are treated, or, in the case of HIV, without using protection, such as condoms.

Waterborne diseases are spread through contact with contaminated water, usually by drinking it, where safe drinking water sources are unavailable. Many infections spread in this way overlap as faecal–oral transmission, where an infection is present in a person's faeces and enters water sources through, for example, poor sanitation or sewage infrastructure, or by such infrastructure being overwhelmed by natural disasters and extreme weather, such as flooding. In some communities, people may defecate directly into water sources if other options are unavailable. Diseases that overlap in this way include diarrheal diseases such as cholera, giardia, and rotavirus, as well as hepatitis A, polio, typhoid and many intestinal worms.

Some waterborne diseases are not spread through faeces but instead through other means of infected humans contaminating water supplies. The parasitic disease Guinea worm, for example, results in worms bursting out of the skin of an infected person, typically their legs. As people soothe their sores in water, the worms lay their larvae which are consumed by water fleas present in the water, in turn consumed by humans who then drink the water, with the larvae growing into adult worms inside the human body. Conversely, some faecal–oral infections, like polio and cholera, can also spread through means other than water, such as by people consuming foods contaminated by unwashed hands, direct contact with the unwashed hands of an infected person, consumption of fish and shellfish that have been swimming in water contaminated with infected faeces, or contact with soil that contains worms from faeces, such as hookworm.

The majority of waterborne and faecal–oral infections continue to occur in developing countries where effective water and sanitation programmes, as well as strong emergency responses, are needed to reduce the problem. Research shows climate change is already exacerbating the problem and will continue to do so as droughts and flooding increase in frequency and impact water sources and availability.

Control strategies become less straightforward when animals are involved in the spread of an infectious disease. The animals bring a new dimension to transmission as they introduce an extra stage to the process and also provide a greater range for the disease due to their mobility. As noted previously, animals involved in disease transmission between people are called vectors; examples include mosquitoes, ticks, fleas, sandflies, mites, and rats. Vector-borne diseases account for 17 per cent of all infectious diseases worldwide, according to the World Health Organization, with the majority spread by bloodsucking insects. The most versatile vector is the mosquito, involved in the spread of multiple diseases, including malaria, dengue fever, yellow fever, Zika and West Nile virus. The life cycle and behaviour of the vector, as well as how the parasite morphs inside it, must be understood to prevent the disease from proliferating.

In many cases, vector control is the best hope public health teams have for disease control and prevention. Insecticide-treated bed nets and the elimination of mosquito breeding grounds, for example, have been extremely effective in reducing malaria transmission in high-burden regions such as Asia and sub-Saharan Africa. Mosquito control, of course, can also reduce the burden of the many other diseases transmitted by that insect. Infectious agents, such as malaria parasites, often have a particular life cycle inside their vector and a different life cycle inside humans, in order to thrive in both hosts and be continuously transmitted. It is through understanding the stages of each cycle that researchers hope to develop vaccines, blocking the parasites' survival.

With some diseases, animals can also become infected themselves, and human encounters with infected animals can often be the cause of emerging disease outbreaks. Animals can also act as reservoirs of a disease that has been controlled or where control is being attempted, often enabling the disease to re-enter human populations. Common animal reservoirs include bats, rats, dogs and monkeys. As noted previously, wild bats, as well as monkeys, are believed to be sources of the Ebola virus passing the infection on to humans who come into contact with them in the wild. Dogs are well-known reservoirs of rabies and Lassa haemorrhagic fever, a disease native to West Africa. Understanding animal reservoirs and either culling animals or limiting contact between animals and humans through control efforts are important aspects of disease prevention, particularly in disease elimination and eradication programmes.

Less common, though a significant concern, is the spread of infections between mothers and their babies during pregnancy, childbirth and breastfeeding after birth. Infections that can spread in this way include HIV, as the virus can transmit through blood during childbirth and be present in breast milk. Antiretroviral treatment significantly reduces the risk of transmission. Syphilis can also spread from mother to baby during pregnancy and childbirth if not treated and hepatitis B can spread during childbirth, with babies needing vaccination soon after birth to protect them.

The struggle to take control

'Everything is about speed, how fast you do it,' says Professor Yuen Kwok-yung, Chair of Infectious Diseases at Hong Kong University, referring to the control of infectious diseases, particularly new or emerging infections. As we heard in the Introduction, Yuen was among the first physicians and researchers to respond to the arrival of SARS in Hong Kong in 2003, having been tasked with understanding this unknown virus and using any insight to curtail its spread. The virus had been spreading in certain parts of China a few months before, with the index case in Hong Kong arriving there from mainland China, but China had not reported the severity of the situation to global health officials and agencies, with some criticising them for covering up the problem. 'If the Chinese government had announced [it] to the world three months earlier, the story may be very different,' explains

Yuen. 'Things need to be controlled as early as possible.' Learning from this, government officials were more transparent as the novel coronavirus behind COVID-19 emerged in the city of Wuhan in 2019/20, though in this case regional authorities were criticized for silencing those that raised preliminary concerns and there have been wider concerns around the accurate reporting of case numbers.

Once authorities are aware, the ideal control scenario involves rapid diagnostic tests, good drugs and a vaccine platform if possible, Yuen believes, though the latter is often unlikely and in some cases all three may be unavailable. In 2017, the Coalition for Epidemic Preparedness Innovation (CEPI) launched to help with vaccine development, financing and coordinating the development of vaccines against known and unknown infectious diseases, with current targets being MERS, Lassa Fever, Nipah, Rift Valley Fever, chikungunya and the as-yet-unknown 'Disease X', which now includes COVID-19. But the road is long and complex and in the meantime outbreaks will continue to occur, surprise us, and spread.

'The most important thing is to deal with an outbreak through the practices we know already,' says David Heymann of the London School of Hygiene and Tropical Medicine. 'There's no rapid cure except good epidemiological practice, outbreak containment' – those practices being contact tracing, isolation of cases, monitoring the people who had contact, and keeping people from travelling into areas where the infection is spreading. In 2020 this has been exemplified by COVID-19, with countries instigating significant measures to prevent further spread of the novel coronavirus, including blocking flights, city-wide lockdowns, quarantine aboard cruise ships and in hotels, and widespread advocacy for self-isolation and good basic hygiene practices, such as regular hand washing.

But Heymann hopes that in the future there will be also be the technology for a vaccine 'backbone' that could have different pathogens slipped in to create a vaccine for different diseases immediately. 'Though even that would take many months and the outbreak will have spread,' he explains, pointing out that use of conventional practices is 'how we stopped SARS'. But in a globalized world, microorganisms are not limited to striking villages, cities or even countries. They now have a whole planet of susceptible bodies that are reachable and while officials in places experienced in outbreaks may know what to do to contain one quickly, once infections reach places that have never faced an outbreak, they can spread rapidly while a plan is put in place.

A prime example is Ebola, which has caused tens of known outbreaks in the Democratic Republic of Congo (formerly Zaire) for over 40 years since its discovery in 1976, each of which has typically been controlled within a few months. But at the end of 2013 the disease struck in Guinea and throughout 2014 and 2015 spread across West Africa, predominantly in Guinea, Liberia and Sierra Leone, where they had never faced the virus before and had very mobile populations. The same applied in 2018 when Ebola began spreading in an eastern region of Congo, North Kivu, which had similarly never seen the virus in its communities before, resulting in an outbreak spanning almost two years. Both outbreaks became public health emergencies of international concern, despite over 20 outbreaks having been contained with relative ease in previous years.

Current concerns also lie around the anti-vaccine movement that has taken root in recent years (more on this in chapter 2), resulting in continuous declines in vaccine coverage for diseases such as measles in developed countries including the United States and United Kingdom. This has resulted in a surge in outbreaks. Estimates suggest that in 2018 more than 140,000 people died of measles – a vaccine-preventable disease – making the disease 'a staggering global challenge' in the words of the World Health Organization. One county in the US declared a state of emergency in 2019, while the United States as a whole saw more than 1,280 measles cases that year, compared to 375 cases in 2018. 'It's a public health challenge,' says David Heymann.

On top of this is growing fear around antimicrobial resistance: disease pathogens becoming resistant to the common drugs used against them, rendering the drugs useless and enabling infections to surge. More on this in chapter 6, but some form of resistant infection has been reported in every country in the world, according to the WHO. Examples include tuberculosis, syphilis, gonorrhoea and many hospital-acquired infections, such as *Staphylococcus aureus*, some of which have become resistant to multiple drugs used to treat them. Despite this, the problem is not getting the global attention it deserves, believes Heymann. A resistant case can travel from any country to any country at any time, he warns – though this applies across the board, be it to a resistant case, a case of a new or unknown disease, an unimmunized individual harbouring a vaccine-preventable infection or someone with malaria or dengue fever arriving in a country where the same mosquito vector can be found.

DISEASE AND POLITICS 2

Infectious disease control, whether relating to a new outbreak or an ongoing epidemic, involves a complex web of people: healthcare workers, physicians, epidemiologists, microbiologists, communication officers, public health experts, community or regional leaders, health ministries or agencies, government leaders and, often, international agencies such as the International Red Cross or Médecins sans Frontières (MSF), as well as UN agencies like the World Health Organization. This is a basic summary, with more people and teams likely to be involved in practice, but it highlights the point – there are a *lot* of people to coordinate, and with different people, environments and cultures involved, things are bound to get political.

In an ideal scenario, invested parties work seamlessly together to provide an effortless, quick and effective response. But in reality, things can quickly get messy. Miscommunication, misalignment and therefore misunderstanding can easily occur, paving a longer road to success. When political instability or conflicts come into play, for example in Yemen or the Democratic Republic of Congo, the road gets even longer. In some situations, an outbreak response may be used as a political tool, to garner support or deflect from other issues a country may be experiencing. In other situations, an outbreak may enable leaders to highlight all the issues a country is facing while the world is watching and in a position to respond. Needless to say, outbreaks inevitably become part of a political agenda.

Take the 2018 Ebola outbreak in the Democratic Republic of Congo, which was ongoing at the time of writing, though new cases had not been reported for two weeks, showing promise of the outbreak potentially coming to an end. But there had been more than 3,440 cases reported as of the beginning of March 2020, including more than 2,260 deaths, the outbreak having begun in North Kivu, in the north-east of the country near the border with Uganda where there is a refusal among some locals to believe the disease exists. People instead support various narratives, including a conspiracy brought in by politicians of the virus being used to target people in the region who have long felt victimised. These beliefs were propagated by the postponement of national elections in the city of Beni and its surrounding areas in December 2018, where people were told they would vote in March due to the Ebola epidemic, while the rest of the country voted as usual. This decision strengthened beliefs that the disease was a political tool and resulted in widespread protests, violence against healthcare workers and vandalism of health facilities, significantly impacting Ebola response efforts. Continued violence has affected the response throughout the epidemic, with a WHO doctor shot dead in April 2019 and health workers under constant security threat. The country has had nine prior Ebola epidemics in other regions, most of which have not escalated beyond a few tens of cases, with just two reaching more than 300 cases. The current outbreak has seen at least ten times this number, fuelled by political unrest.

Ebola can also be used to highlight the influence of politics more globally on outbreak control efforts, if we consider the example of the 2014–15 outbreak in West Africa, which overshadows the Congo outbreak having caused more than 28,000 cases and 11,000 deaths in the region. Many experts describe this outbreak as involving the 'perfect storm' of factors to fuel a disaster, most of which were political. When Ebola arrived in Guinea, porous borders helped the infection spread with ease among unsuspecting, mobile populations to Liberia and Sierra Leone. Government officials had never seen the disease or faced this kind of situation and health teams worked in a system and overall infrastructure damaged by years of civil war, while needing to calm down panicked communities that had little trust in their leaders. Then came the international agencies, with resources and skilled teams to help treat cases and block further transmission. The agencies worked with communities where there was a need to build a rapport, making this their first objective, while sending word back to the rest of the world that things were under control – or at least could be, if countries pooled together to supply the funds, technologies and supplies needed. The World Health Organization was hugely criticized for downplaying the scale of the situation at first and failing to act quickly enough with its guidance, allowing the outbreak to surge before declaring a public health emergency of international concern. Some argue the complexity of this outbreak, and the failure to respond quickly, were why the outbreak spanned two years, rather than a matter of weeks or months, as had been the case for prior Ebola outbreaks.

But raising the alarm early is not always the right thing to do, as it could be raised too soon, eliciting unnecessary panic or, even worse, wasting national and global resources. This was the criticism five years earlier of the response to the 2009 H1N1 influenza pandemic, despite the virus reaching 214 countries and killing more than 18,500 people. The Council of Europe, an international organization that protects human rights and the rule of law in Europe, heavily criticised the WHO as well as national governments and European Union agencies for their handling of the situation, concluding that money and resources were wasted and fear brought upon the public without adequate justification. The virus turned out to be milder than originally thought, infecting fewer people than first predicted. This was in part due to the discovery that some older populations were immune as the virus had similarities with one that circulated in the 1960s, leaving those exposed at that time with some immunity to the new strain.

The council believed that the WHO had not been transparent enough about the backgrounds of the people on the emergency committee tasked with assessing the severity of the pandemic, believing some had ties to pharmaceutical companies that stood to gain from stockpiling drugs and vaccines. The WHO defended its decision, denying links to the pharmaceutical industry and highlighting that early research and epidemiology suggested they were dealing with a new, unique virus with moderate severity but little immunity among populations. Some experts argue agencies must be prepared for any impending pandemic, while others argue the need to hold the trigger if billions of dollars are to be wasted. Either way, it's far from a straightforward situation and, inevitably, someone is likely to be criticized.

The power of anti-vaxxers

The anti-vaxxer, a person who opposes vaccination on various grounds, is a term most people are now familiar with, along with the wider anti-vaccination movement, the modern version of which promotes misinformation, myths and conspiracy theories about immunizations in a bid to turn people against them. Today's anti-vaxxers endorse vaccine hesitancy, the reluctance or refusal to vaccinate, citing a diverse range of reasons using multiple platforms, depending on their audience, creating a complex obstacle for health officials to overcome.

The movement itself has been around for centuries, stemming as far back as the first vaccinations against smallpox. In the nineteenth century, posters were often seen telling people not to get vaccinated because of the associated dangers. But the modern voices of this minority are louder, stronger and now digital, recruiting new members in greater swathes and causing the return of vaccine-preventable diseases such as measles, which, as we have seen, caused unexpected outbreaks in almost every region of the world in 2018 and 2019. Globally, there has been a 30 per cent rise in measles cases, with an estimated 140,000 deaths in 2018, and although the reasons are complex, most experts believe vaccine hesitancy is in large part to blame.

It's a global phenomenon, playing out differently depending on the setting, says Dr Heidi Larson, Director of the Vaccine Confidence Project at the London School of Hygiene and Tropical Medicine. Vaccine hesitancy has now been reported in more than 90 per cent of countries in the world, with many now seeing less than 95 per cent of the population protected against measles. Larson's project has been monitoring public confidence in immunization programmes globally for ten years, developing a vaccine confidence index to map public opinion by country and region. The team define confidence as the trust in the effectiveness and safety of vaccines and trust in the healthcare system that delivers them. A global survey by the project in 2016, across 67 countries, found developing countries such as Bangladesh and Ethiopia had the most confidence in vaccines, while Europe had the lowest confidence of any region, particularly with regard to the safety of vaccines: seven European countries were among the ten least confident countries worldwide. Highlighting the shift in modern society was the finding that countries with higher levels of schooling and better access to health services were associated with lower positive sentiment towards the use of vaccines. A deeper look into attitudes in the European Union in 2018 found that many countries had become even less confident in the safety of vaccines by then, including Czech Republic, Finland, Poland, and Sweden, often correlating with general practitioners in those countries questioning the safety and importance of vaccines.

A 2018 global survey by the Wellcome Trust further validated this trend, finding that just 59 per cent of people in Western Europe and 50 per cent of people in Eastern Europe agreed vaccines are safe – compared to 95 per cent in South Asia and 92 per cent in Eastern Africa. The United States, though it had higher confidence figures than many European countries, had the lowest belief in vaccines of any country in the Americas. As a result, people are choosing not to vaccinate their children and diseases like measles, once considered eliminated in some countries, are seeing a surge in outbreaks.

A 2018 report by the Royal Society of Public Health (RSPH) in the UK found that there was a low level of understanding about herd immunity among working-age adults, while views on vaccine overload – the belief that receiving too many vaccines can be harmful – were also found to be persistent, with one in four people believing you can have too many vaccines. This was one argument formerly used by US President Donald Trump, before he took office. On the plus side, 91 per cent of parents believed in the importance of vaccines for children's health – but safety fears were then cited as a reason why some may nevertheless choose not to vaccinate.

The global situation prompted the World Health Organization to declare vaccine hesitancy to be among the top threats to global health in 2019. The global agency stated: 'vaccine misinformation is a major threat to global health that could reverse decades of progress made in tackling preventable diseases.' But what makes people hesitant and, importantly, what makes them believe everyone should be hesitant? Aside from some religious beliefs, which have long been against vaccines in some way, Larson thinks there are three core reasons in modern society: libertarians wanting freedom of choice or wanting to go against their government or authorities; people who want purity, are against chemicals and believe in a natural life; and, most dominantly, people concerned about the safety of vaccines who then spread their anxiety.

Religion remains a common reason communities use to avoid vaccination, with significant impact. A pair of measles outbreaks occurred in New York City and nearby Rockland

County in Autumn 2018 following the return of unvaccinated travellers who had visited Israel, where a measles outbreak was ongoing. The New York/Rockland County outbreak went on for almost one year, causing more than 600 and 300 cases of measles respectively, making it the worst outbreak in New York in almost three decades. The cases were predominantly among the minority ultra-Orthodox Jewish community, in which parents refused to vaccinate their children, whereas in many Orthodox communities religious leaders promote the benefits and importance of vaccination. A state of emergency was declared in the city and vaccination made mandatory in certain areas to try to rescue the situation. The US as a whole has seen a recent surge in measles, with 1,282 cases reported across 31 states in 2019, the greatest number since 1992, leaving it at risk of losing its elimination status. More than 73 per cent of the cases were linked to the outbreaks in New York and the majority were among unvaccinated people. Despite this increase, the country maintained its measles elimination status.

In northern Nigeria and Pakistan, two of just three countries still endemic for polio, efforts to vaccinate communities have seen health workers attacked and parents refusing to vaccinate their children against the disabling disease due to misinformation in the name of religion. In Pakistan, Islamist militants spread fear that immunizations are a Western conspiracy to sterilize Muslims, for example, causing polio to persist in communities despite an effective vaccine being available.

Mothers wanting a 'natural' life for their children share a range of reasons for refusing vaccination, with studies finding beliefs among them that vaccines are 'unnatural' or 'impure', or that they weaken immune systems, meaning healthy children do not need them. This group questions medical intervention as a whole, even during pregnancy, aided by alternative and complementary medical providers, says Larson. Homeopathy has quite a following across Europe and the UK, she explains, and in the case of the 'natural' parents, some believe getting measles will in fact strengthen their children.

However, the biggest vaccine-hesitant group are the parents who fear the safety of vaccines. It is apparent that their beliefs, and the worry and anxiety that accompany them, are somewhat contagious, particularly when spread by notable figures and also now spread with ease using social media. The Wellcome and Vaccine Confidence Project surveys clearly show a rise globally in people questioning vaccine safety. The spread of fear was evident on the Pacific island of Samoa, for example, which saw a significant drop in vaccine coverage, falling as low as 31 per cent, following the deaths of two babies in July 2018 during routine measles vaccination. The fatalities led to strong anti-vaccine sentiment nationwide and eventually an unprecedented surge in measles cases in 2019, with the island nation declaring an outbreak in October, expanding this to a national emergency by November and seeing over 4,800 cases by mid-December, including more than 60 deaths.

Many experts agree that the issue of vaccine hesitancy as a whole, and in particular its scale in relation to the combined measles, mumps and rubella (MMR) vaccine, was rejuvenated by research papers published in 1998 by the now discredited physician Andrew Wakefield, who suggested the MMR vaccine might cause autism.

Wakefield drew extensive criticism from peers in the field for the 'flawed and unethical research methods' he used to gather his data and draw his conclusions, which included small sample sizes and incomplete links between his data findings and conclusions. Investigations also revealed conflicts of interest as he had received funding from litigants against vaccine manufacturers. This all led to the journal he had published the findings in, the *Lancet*, retracting his paper and the General Medical Council striking Wakefield from the UK medical registry, barring him from practising medicine.

Subsequent studies have continually disproven any link between the MMR vaccine and autism, with yet another published in 2019 by Danish researchers studying more than 650,000 children and finding that 'MMR vaccination does not increase the risk for autism, does not trigger autism in susceptible children, and is not associated with clustering of autism cases after vaccination'. This has been the verdict time and time again, with each study growing larger in size, giving the findings more strength, but the damage was done by that first paper, particularly in Europe and North America, and continues to thrive in groups who remain blind to anything proving them wrong. The time taken to retract the paper, twelve years, also gave enough time for the findings to go deep, says Larson.

In the UK, MMR vaccination rates dropped from 92 per cent in 1996 to 84 per cent in 2002. In Ireland the national rate fell below 80 per cent by 2000 and the rate in the US that same year was 90.5 per cent. Unsurprisingly, measles outbreaks soon occurred. The UK saw 56 measles cases in 1998, but this went up to 449 in 2006 and in 2018 there were more than 910 cases in England alone, with the UK losing its official status of 'measles free' in 2019. Public Health England called repeatedly for all parents to have their children vaccinated.

Larson believes that the impact of Wakefield's paper was enhanced by factors both intrinsic and extrinsic. 'People forget his paper came out the same year Google opened its doors, followed by Facebook, Twitter, Instagram and Youtube,' she says. In that respect, 'he got lucky', she says, adding that Wakefield also touched a nerve with a very dominant concern among parents: autism. He gave them a cause for a condition they were affected by or worried about. This is also why papers disproving his findings haven't been shared as widely, she believes, as no alternative was given in its place. 'To just say it's wrong without coming up with something better, is a non-starter,' says Larson. 'No one has given parents a better answer.' Since then, Wakefield has built a campaign, a movement, and been welcomed into the libertarian and alternative therapy groups by 'saying all the right things'. In other words, he hasn't gone away and it doesn't seem like he has any plans to. 'He's still highly active and very well funded,' says Larson.

The rise of technology and, ultimately, social media has played a key role in the growth of anti-vaccine sentiment, via online groups and message boards spreading conspiracy theories, misinformation and generally providing a forum for people to validate their views against vaccines. But controlling this is complicated, as with anything relating to the tech sphere. Social media companies have been called out and urged to address the concerns around the role their platforms are playing in the anti-vaccination movement, particularly fearmongering, and instead to promote the spread of accurate information and links to trustworthy sources. The RSPH report identified social media as a key tool in propagating negative messages around vaccines in the UK, particularly among parents. Two in five parents said they were exposed to negative messages, with this proportion going up to one in two among parents with children under five years old. The health agency called for social media companies to be held to account.

In September 2019, Facebook announced that educational pop-up windows would appear on Instagram and Facebook whenever users searched for vaccine-related content, linking to credible sources, such as the World Health Organization. This earned some praise from health experts, though some, including the British Medical Association, believe anti-vaccination content should be banned altogether. Earlier that same year, Instagram blocked the use of anti-vaxxer hashtags, such as #vaccinescauseautism and #vaccinesarepoison and vowed to stop other hashtags caught being used to spread anti-vaccination sentiment. But numerous accounts held by anti-vaxxers, even with the word in their username, persisted (and still do), with the company saying they could not take action against the accounts of people who identify as anti-vaxxers. As companies develop ways to prevent the spread of fake news and information, those in the movement find caveats and loopholes to continue spreading it. 'Most of them are clever enough to not say anything that would be illegal,' says Larson. 'They're posing questions, putting billboards up, saying, "Do you know what's in a vaccine?" or "Do you really know the risks?".' With this comes the use of hashtags including #informedconsent and #runtherisk, which again straddle these loopholes. Tech companies and governments cannot shut down the doubt instilled in the public through simply asking questions, which Larson stresses as the bigger challenge. 'They're being much more responsive than official sources, constantly alert to the environment and changing their narrative where they see opportunity,' she says. In contrast, the official voices and the pro-vaccine groups are not being responsive enough and do not understand the social media sphere, Larson believes, and this is to their detriment.

Research within Larson's group by Dr Sam Martin explored social media discourse in late 2018 around the use of vaccines during pregnancy on a global scale to see which regions and countries had the most positive and negative views towards them. Across fifteen countries, in all regions, most social media posts analysed were positive towards the use of vaccines during pregnancy but significant amounts of negativity were found in Italy and the US. Negative content suggested risks surrounding the influenza vaccine and linked vaccines to different diseases, including Zika and autism, using misinterpreted scientific evidence to aid the claims. Some posts in the US also referred to protecting children from vaccines rather than from disease, as well as mistrust towards health institutions such as the Centers for Disease Control and Prevention. Martin's studies also showed the ease of information spreading across countries using social media platforms. 'We shouldn't think issues in London stay in London,' says Larson.

To beat the movement, Larson believes authorities must implement a range of solutions. She highlights the issue of missed vaccinations, due to people either not showing up or simply not getting an appointment in the first place. Millions of appointments are missed in London alone, she says, which easier access to vaccinations and greater follow-up programmes would address. Group behaviour and dynamics also need to be understood, she believes, to see how people influence one another, and health authorities need to enter social media and other places that involve emotion, as 'that's where people are'. At present, says Larson, the health authorities are 'not getting into the conversation'. It's clear that the questioners and the doubters are quick to reach out to get people on board with their views, which is what those endorsing vaccines need to catch up with in order to regain their power.

Next on the manifesto: vaccines

The uptake of vaccines and beliefs around their benefits or harms are in no small part linked to a person's political beliefs and by extension those of their political leaders. Notably, when leaders are vocal against vaccination, they undoubtedly boost the anti-vaccine cause by giving a platform to people desperate to spread their views. One example is US President Donald Trump who, prior to being elected President, made multiple false claims about vaccines causing autism, many of which were on Twitter. 'Massive combined inoculations to small children is the cause for big increase in autism,' he tweeted in 2012; and in 2014: 'Healthy young child goes to doctor, gets pumped with massive shot of many vaccines, doesn't feel good and changes – AUTISM. Many such cases!' He repeatedly made similar comments on the record and was seen publicly with high profile anti-vaxxers including Andrew Wakefield and Robert Kennedy Jr, giving greater power to those who shared his viewpoint.

During Trump's presidency, the US has seen a surge in measles cases, reaching record numbers since the disease was eliminated in 2000. The urgency of the situation caused a switch in his stance, with Trump telling parents they 'have to get the shots' in April 2019, but it could be argued that years of speaking against them gave confidence to those refusing the vaccines, inevitably resulting in outbreaks. Anti-vaxxers 'were emboldened to think they're right', and Trump's tweets and comments continued to have traction once he was in office, says Dr Heidi Larson.

Despite Trump and the US, Larson's global surveys found that Europe had the most anti-vaccine sentiment of all regions, reflected by surges in measles cases in recent years, with more than 80,000 cases in 2018, increasing to more than 100,000 in 2019. Numbers were a quarter of this in 2017, at just shy of 24,000 cases. Political leadership in certain countries boosted anti-vax rhetoric. Italy, for example, has long had a prominent movement against vaccines, highlighted by measles vaccine coverage dropping to 85 per cent in 2015 and resulting in catastrophic outbreaks in 2017 that caused more than 5,000 cases, with cases staying above 2,000 in 2018. The country's anti-establishment Five Star Movement, now in power, tapped into this sentiment and the country's general distrust of institutions, including the medical establishment, to gain favour by echoing the views of libertarians who believe vaccines are a choice that can be refused. A distrust of elites and experts was propagated, in line with the general shift towards populism across Europe and more globally.

A 2019 study analysing voting data and views on vaccines in Western Europe found a significant link between populism and vaccine hesitancy. The study, led by Dr Jonathan Kennedy at Queen Mary's University of London, found 'a highly significant positive association between the percentage of people in a country who voted for populist parties and who believe that vaccines are not important'. The association with beliefs that vaccines are not effective was also significant. The countries with the greatest link between populist support and disbelief of vaccines were France, Italy and Greece, all of which have seen extensive measles outbreaks in the past few years, the worst by far in the European Union in 2018, with more than 2,000 cases in each country.

A 2019 survey across 24 countries by the YouGov-Cambridge Globalism Project found similar associations, with people holding populist beliefs being more likely to believe in conspiracy theories that go against science. People with strong populist views were almost twice as likely to believe that the harmful effects of vaccines were being hidden from the general public. The same survey further supported insight about people who vote for Italy's Five Star Movement, finding they were more likely than other voters to doubt scientific evidence. The evidence is building up to paint the reality of the political world today: an increasingly disenfranchised population pushing back against the elite, the experts, in multiple ways, including the refusal to vaccinate their children out of fear that the truth in general is being withheld.

In the face of this, however, are political leaders – many of whom are leading or in coalition with right-wing parties – and health officials witnessing measles outbreaks in their country. This reality caused President Donald Trump to do a public U-turn on the matter, and more recently Italy's Five Star Movement did not follow through on promises to scrap laws requiring parents to prove their children had received ten mandatory vaccinations, including against measles, in order to attend nurseries and primary schools. The Italian law has been a subject of controversy since its inception in 2017 following the rise in measles cases, with those in favour highlighting the rise in vaccination coverage since it was introduced and those against arguing it alienates people. In March 2019, a number of children were turned away from school gates across the country for failing to show their vaccination certificates.

Similar debates about mandatory vaccination occur across the United States, where all states require children to receive certain vaccinations, including measles, to attend public school but many have different exemptions, including medical exemptions and religious and personal beliefs. Just five states allow solely medical exemptions, including California and New York, which both introduced bans on personal belief exemptions following large outbreaks of measles – California in 2015 and New York in 2019. In 2019, California further tightened its rules after the state saw a rise in medical exemptions from parents using this route as a loophole. A standardized exemption form was introduced, along with a directive that doctors providing more than five exemptions a year would be investigated, as would schools showing vaccination coverage below 95 per cent. Both states have since seen a rise in home schooling.

Australia has even stricter measures in place, requiring parents to have vaccinated their children in order to receive childcare and maternity payments, with the government stating 'the choice made by families not to immunize their children is not supported by public policy or medical research nor should such action be supported by taxpayers in the form of child care payments'. More recently, some states in Australia have added the requirement to be vaccinated in order to attend preschools and childcare centres. In both instances (payments and childcare) only medical exemptions are allowed. Similar discussions are taking place across the Western world, with France also making ten vaccines mandatory among children in 2017 and Germany now requiring kindergartens to notify authorities of parents who haven't at least had vaccination counselling for their children, to explain the benefits. One thing all countries have in common is the pushback received from anti-vaxxers, including death threats faced by those implementing the new laws and measures.

Larson believes mandatory vaccination isn't necessarily the right thing to do for all vaccines, but with measles being the most infectious of all vaccine-preventable diseases, and having serious consequences from infection, it's reasonable to consider in settings like schools and hospitals. Many experts, however, including Larson, argue that issues around vaccine access are just as important, if not more so, with resources needed to engage and follow-up with parents, prompt and remind those who are behind on their vaccinations and also ensure vaccine supplies are adequate so people get them on time and as needed. For example, some parents don't vaccinate simply because they have fallen behind with life admin and others who are anxious may simply need to talk to health professionals in a dialogue about the benefits rather than in an authoritarian manner. Governments need to make sure that missed vaccination appointments are followed up, and that those trying to book appointments get them, with these factors being particular concerns in the UK, says Larson. The RSPH report found that 'improving access to vaccinations remains crucial especially when tackling inequalities in uptake, for example relating to ethnicity or socioeconomic status'. Larson adds, 'You can't impose a mandate if you're not ready to deliver.'. The Australian government also allocated budget for immunization awareness campaigns, deemed by many to be crucial in the age of easy and endless access to misinformation.

Whether mandatory vaccination is effective has long been debated, with many experts arguing each side. But Australia has seen an increase in immunization rates since introducing its measures, as have California, France and Italy. Some argue that the problem of anti-vaxxers is not best overcome by forcing them to vaccinate, but to instead engage in a conversation where you identify with them and acknowledge that the science around vaccines can be imperfect, but that the benefits far outweigh the risks, hence making them a good option to choose.

Others believe the issue comes down to the fact that the people choosing not to vaccinate are the people who have not borne witness to the ravages of deadly disease. In an opinion piece, Professor Tom Solomon from the Institute of Infection and Global Health at the University of Liverpool argues that 'the whimsical pontification on the extremely small risks of vaccination' is a luxury only available to people in the developed West, where diseases like measles and polio are rarely seen. Solomon explains how families in Asia queue for hours to receive vaccines from the programmes he works on to protect their children from the horror of diseases they have witnessed all too often. When asking whether vaccines should be made compulsory, Solomon gives the examples of numerous precedents that have paved the way for this not to be such an infringing option – proof of yellow fever vaccination for travellers from certain countries in Africa and South America, courts overruling parents refusing life-saving medicines for various reasons including religious beliefs, and even seatbelts being compulsory in cars, despite them potentially causing harm in rare circumstances. 'Faced with the horrors of the diseases they prevent, most people would soon change their minds,' he writes in the *Conversation*.

The movement against vaccines is unlikely to disappear, but in the light of preventable diseases re-emerging and a large part of this coming down to the spread of false information, feeding into already doubtful or religious minds, political leaders and tech giants need to face the reality of the situation and continue to do what's needed to protect the population as a whole. What that may be will change as society continues to evolve.

LONG LIVE DISEASE 3

When thinking about infections, particularly ones that cause outbreaks and epidemics, it's easy to just consider the new, dramatic diseases that appear on the news, putting the world into a state of flux. But a handful of diseases we see today have been around for centuries – some for millennia – and while their presence has slowly reduced over the years, some show no signs of truly going away.

Most of the ancient diseases that continue to plague communities today are found in developing countries – typically in rural areas of extreme poverty – and are often therefore not on the radar of those living in the developed West, going unreported in the mainstream media until they arrive closer to home. Take leprosy as an example: a disease many associate with biblical tales, ancient texts or historical novels, telling of people stigmatized and ostracized from society, perhaps shipped off to islands to live in colonies away from the rest of the world. But the disease is very much present today, with more than 208,000 new cases registered globally in 2018, across 127 countries. Leprosy was targeted for elimination as a public health problem, which meant bringing prevalence of the disease down to less than one case per 10,000 people, and this was achieved in 2000. Case numbers today are approximately 0.2 per 10,000 people. But with a global population of almost 8 billion people, that leaves a lot of cases, though 79.6 per cent of them occur in three countries: India, Brazil and Indonesia. In 2016, the WHO restarted efforts for a 'leprosy-free world', focusing on reducing infections among children, as the bacterial infection is curable if caught early with a course of three antibiotics. Many myths and uncertainties remain around the disease, such as the belief leprosy is spread by casual contact – simply touching an affected person. More recent evidence has been in favour of it being spread through respiratory transmission, via air droplets coughed or sneezed by untreated patients, and through prolonged contact. Infection with the bacteria, *Mycobacterium leprae*, causes progressive damage to the skin, nerves, limbs and eyes if left untreated. The WHO provides treatment for free through an agreement with pharmaceutical company Novartis, but the main barrier today is finding those affected in time.

Leprosy belongs to a group of diseases referred to as 'Neglected Tropical Diseases', a diverse range of infectious diseases that occur in tropical and subtropical regions, typically affecting the poorest communities who lack adequate access to water, sanitation and basic health services and who are also most exposed to disease vectors such as mosquitoes. Other diseases in this group include intestinal worms, African sleeping sickness, rabies and lymphatic filariasis, also known as elephantiasis – historic diseases that unfortunately persist today.

Plague is another disease people may defer to biblical texts to read about rather than present-day outbreak reports. But the disease, once called the Black Death, continues to have a presence, albeit a much smaller one than leprosy, with 3,248 cases reported globally between 2010 and 2015 according to the WHO. An unusual outbreak in Madagascar, where the disease is endemic, in 2017 saw it reach unprecedented scales, with more than 2,300 suspected cases and more than 200 deaths over a four-month period from August to November. Once causing 50 million deaths in Europe alone during the fourteenth century, plague today is endemic in three countries: Madagascar, Democratic Republic of Congo and Peru. But sporadic cases occur elsewhere – for example, two cases were reported in China in 2019 and health officials raced to stop a larger outbreak from happening.

Plague is a bacterial infection spread through the bite of infected fleas or contact with infected animal tissue or bodily fluids and is severe if left untreated. The disease occurs in two forms: bubonic plague, causing inflamed lymph nodes, and pneumonic plague, a more severe form infecting the lungs, which can then spread from person to person through air droplets. Both forms are, however, curable with antibiotics – they just need to be caught in time. This marks a trend among many of the ancient infections that persist today; they lurk in the shadows, hiding among the most vulnerable, unlikely to be seen by health services in time to treat those affected and stop the infection spreading. Through this secretive existence they may continue to thrive for many more centuries – outliving the people who forgot their existence.

Tuberculosis: what happens when nobody cares

As diseases go, it could be said that tuberculosis (TB) is up there among the most common, the most complex, the most perplexing and, to humanity's detriment, the most forgotten. The disease is one of the top ten causes of death worldwide and is the world's biggest infectious killer – ranking above HIV/AIDS – yet, it's fair to say that many people, typically in the developed world, are surprised to hear tuberculosis is still around at all, let alone to learn that 10 million people fell ill with the disease in 2018, almost 2 million of whom died.

Its complexity lies in the biology of the disease, caused by the bacterium *Mycobacterium tuberculosis*, which most often affects the lungs (this is called pulmonary tuberculosis), causing the coughing up of sputum and blood, as well as weight loss and night sweats. In more rare circumstances, the bacteria can infect the lymph nodes, bones and joints, digestive system, bladder, reproductive system, and even the brain and nervous system, to cause extra-pulmonary TB (TB outside of the lungs) and even cause a form of meningitis. The bacteria also have the ability to use a person's own immune system against them to form a granuloma, an organized array of immune cells, inside which the bacteria can hide away, leaving them unrecognized by the wider immune system and the person symptom-free indefinitely, until something, such as HIV or diabetes, causes their immunity to weaken and the bacteria to break free and attack. As a result, it's estimated that just 5–10 per cent of people infected with TB will go on to develop the disease, with most doing so within two years of contracting the infection. It's further estimated that a quarter of the world's population is latently infected.

Perplexity surrounds the fact that TB continues to exist on a large scale today, despite the bacteria not being particularly contagious, the disease being treatable with a regimen of antibiotics, and it being preventable both through basic hygiene practices and by pre-emptively treating those with latent TB. The reason it persists is explained in some way by the fact TB is often forgotten about: dismissed as a low priority, enabling it to sneak back into communities who aren't watching it closely.

It's a disease of poverty, a social disease with medical implications, explains Dr Paula I. Fujiwara, Scientific Director of the International Union against Lung disease and Tuberculosis, often referred to as the Union. 'Do people care about people that are poor? Not always,' she says. When asked why TB is still around, Fujiwara opines it comes down to a genuine lack of attention: health officials, government departments and funders simply not worrying enough about the disease to prioritize it over others. 'There was no TB programme at WHO until 1992,' she points out. 'The Union was taking on the function of a WHO.' Fujiwara explains that a hundred years ago there was a social movement, demanding treatment and action to end tuberculosis. This saw the first anti-tuberculosis drugs materialize in the late 1940s and 1950s: streptomycin, then isoniazid, the latter still used in drug regimens today. But while the drugs left patients dancing with glee in the aisles of sanatoriums, Fujiwara explains they were both a blessing and a curse: 'Because it became something that was treatable, people said, "You don't have to worry about it anymore."'

The challenge that came with TB treatment – a challenge that continues to be the backbone of problems around ending the disease today – was its duration. Completely curing someone of tuberculosis required him or her to take a set regimen of antibiotics every day for at least six months. Failure to complete the course would mean the bacteria were not fully wiped out and, as is the case now, the remaining bacteria could sometimes evolve to develop resistance against the drugs being used to kill them. In 2018, around

500,000 people with drug-resistant TB were reported, with 78 per cent of them resistant to more than two drugs used to treat the disease. The issue of resistance, and failed treatments, first arose from a lack of follow-up procedures once people were diagnosed with TB and prescribed treatment: health workers wouldn't know if patients were taking the drugs regularly or not. In 1993, the WHO declared TB to be a global emergency and two years later launched a strategy called Directly Observed Treatment (DOT), which, as well as ensuring commitment, diagnostics, drug supplies and surveillance resources, would require patients to present at health facilities regularly, in many cases daily, for health practitioners to witness them taking their drugs. 'There were no new drugs coming, so we needed to protect the drugs we had by making sure people took their medication and prevented resistance,' explains Fujiwara. They needed to 'make sure that everyone that started the medication actually completed it'.

Tuberculosis rates have come down since the emergency was declared, but not enough, and rates of new infections today are declining by just 2 per cent annually. A goal has been set to eliminate the disease in 30 countries by 2050 – elimination being defined in this case as less than 1 case per million people globally – and a global strategy to end the TB epidemic hopes to reduce TB deaths by 95 per cent and to cut new cases by 90 per cent by 2035. But if things continue at their current rate, neither of these will happen.

Some lessons could be learned from Fujiwara's time in New York in the early 1990s, where she helped U-turn a city at crisis point. Recruited by the US Centers for Disease Control and Prevention, Fujiwara went to New York in 1992 when the city was seeing extensive outbreaks of TB. Rates of the disease had been rising throughout the city in a multi-drug-resistant,

HIV-related epidemic with multiple hospital outbreaks. 'This was totally out of control,' Fujiwara explains, because 'no one cared about it', leaving the CDC to intervene and bring Fujiwara over to help lead on control efforts.

Between 1978 and 1992, TB rates had tripled in New York City and the proportion of those who had multi-drug-resistant TB had doubled, the increases fuelled by diminished control efforts, rising poverty and homelessness, and overcrowding. The city had successfully controlled TB rates in the 1960s but increasing success led to decreasing resources as officials believed the situation was under control. 'By 1992, the situation in New York City looked bleak,' Fujiwara and her team wrote in their paper, with 3,811 cases of the disease recorded and more than 20 per cent of those having a drug-resistant form of the disease. Experts believed failed treatment compliance was at the heart of the epidemic of drug resistance. In one of the city's poorest neighbourhoods, Harlem, the TB infection rate of 222 cases per 100,000 people was higher than in many developing countries at the time. 'There was no real programme to address these people,' says Fujiwara. So her team set out to introduce one, focused on preventing the spread of the disease, largely by ensuring people completed their treatment and, among officials, using the mantra 'Think TB'. 'No one was thinking about it at that point, it had gone away, wasn't as big a deal as before, but it sneaked back in.'

In their research paper, Fujiwara's team outline three core control measures they prioritized to bring down the rates of TB in the city. The first was directly observed treatment, which saw treatment completion rates reach 90 per cent among diagnosed patients by 1994. By preventing patients from becoming or remaining infectious, the team calculated that they prevented at least 4,000 TB infections between 1992 and 1994. This strategy has since, however, had pushback from advocacy groups arguing it asks too much of patients, many of whom have jobs they rely heavily on, to present at health facilities on a daily basis. But the number of cases in NYC fell quickly, by almost 1,000, to 2,995 in 1994.

The second control measure prioritized was improved infection control in hospitals, better screening, isolation and follow-up of prisoners and the downsizing of large homeless shelters – as both prisoners and the homeless (more specifically, the poor) see high rates of TB – all helping reduce the spread of TB. The final measure was a new drug regimen, consisting of four antibiotics, to tackle drug-resistant bacteria and shorten the amount of time patients remained infectious. With almost 3,000 cases in 1994, the team had a long way to go to truly tackle the problem, but declines were seen and in 2018 there were 559 verified cases of TB in NYC, equating to 6.8 per 100,000. This is nevertheless more than double the US national rate of 2.8 per 100,000. If we consider that the global rate in 2018 was 130 TB cases per 100,000 and that countries including Namibia, South Africa, Philippines, North Korea and the Central African Republic had rates above 500 per 100,000 that same year, it can be seen that the global burden of TB remains vast.

Strategies to reach the ambitious global targets set by the WHO include addressing the most vulnerable groups, screening for both active and latent TB in high-risk groups, and investing in new tools and diagnostics. Fujiwara further highlights the diagnostics point, describing the need for a point-of-care test that can immediately reveal if someone has TB, like HIV testing, as the main form of diagnosis today remains laboratory analysis of sputum for the presence of bacteria, which can take days or even weeks in the case of drug-resistant infections. But she also emphasizes the need for new, shorter drug regimens, including the use of drugs to treat people at risk and who may have latent TB: people with HIV, for instance, and close contacts of TB patients – the aim being to prevent them developing the disease and spreading it to others. 'We cannot end the TB emergency unless we dramatically scale up prevention in those parts of the world where we are treating it,' she says.

But Fujiwara also has high hopes for a vaccine, describing it as the ultimate prevention tool, with multiple candidate vaccines currently at the research stage and some reaching clinical trials. However, a vaccine is unlikely to have 100 per cent efficacy – more like 50 or 60 per cent. The current Bacillus Calmette–Guérin (BCG) vaccine against TB is mainly given to babies and children at risk of being exposed to the infection, to protect them against the non-pulmonary forms of the disease, which are more common among this age group. It provides little to no protection against the respiratory form of TB in adults – the most common form. But the reality of delivering a new vaccine to all affected populations is unlikely and would again become a tool prioritized for people at risk, Fujiwara explains.

So, to conclude, is this centuries-old disease likely to disappear anytime soon? The answer is most likely no. A lot needs to happen to remove these persistent bacteria both from our environment and from the bodies within which they are hiding, but the targets are set to keep momentum going and in hope of getting those in charge to care – to 'Think TB'. The question is how much they will continue to care when concerns around TB are weighed against everything else they have to worry about.

The persistence of polio

Poliomyelitis, known more commonly as polio, was once a disease feared worldwide, striking anyone in its path but particularly children, often paralysing them for life and killing up to 10 per cent of those infected. The first case was reported in 1789, with numbers rising and the disease spreading in the decades that followed. The virus spreads through food and water contaminated by the faeces of an infected person, and once contracted, the disease has no cure. Polio is now close to being wiped out for good, but has been at this stage for some time, as mistrust, misinformation and multiple conspiracy theories block the final road to total eradication.

The global programme to eradicate polio began in 1988 when there were still an estimated 350,000 cases across 125 countries. There was dispute among experts about whether eradication could be achieved, because, unlike smallpox, not everyone infected by the virus becomes paralysed or symptomatic. Therefore, the strategy had to be different to that employed for smallpox, vaccinating everyone across all populations. But this soon proved to work. By 2000, polio had been eliminated from the Americas and what the WHO designates as the Western Pacific region, followed by the European region in 2002. By the end of 2019, just two countries were endemic for the disease, reporting tens of cases each, and this success in large part comes down to the protective shield the polio vaccine has given to the majority of the world's population. 'Polio was eliminated in most countries very rapidly because they had high levels of routine vaccination coverage,' says David Heymann, Professor of Infectious Disease Epidemiology at the London School of Hygiene and Tropical Medicine. This meant that a few extra vaccination campaigns to boost vaccine coverage were enough to reach the herd immunity threshold and wipe out the disease. To be declared free of polio a country or region needs to have been free of the disease, with no new cases, for three years.

But delivery of the polio vaccine has been shrouded with controversy in the final two countries still harbouring the disease: Pakistan and Afghanistan. Nigeria, which is close to being declared polio-free, having not reported cases for over three years, also saw extreme circumstances on its road to defeating the disease. It first became polio-free in 2015, but two new cases were reported in 2016. Multiple conspiracy theories around the polio vaccine publicized by political and religious leaders across these three countries has led to the regular withdrawal of children from immunization campaigns by their parents, aided by an entrenched mistrust against Western countries and health agencies.

In Nigeria, reports and studies have described this mistrust being predominantly present in the north, home to the country's Muslim population. Things escalated in 2003 when leaders in a handful of northern states endorsed community rumours around the vaccine and told their respective communities that polio vaccines were contaminated, for example with HIV, or chemicals to sterilize young girls, and called on parents not to allow vaccination teams to immunize their children. Immunization campaigns were brought to a halt. 'Vaccination stopped throughout northern Nigeria and polio spread to eighteen countries that were polio-free,' says Heymann, who previously worked on the polio eradication programme at the WHO. Polio cases in Nigeria jumped from 202 in 2002 to more than 1,100 in 2006. Global health teams had to then engage with religious and political leaders to help explain the science behind the vaccines. In this they received support from the Organization of Islamic Cooperation, which helped counter myths that the vaccine was designed to sterilize Muslims. 'It took a year and a half to get things back on target again,' Heymann says.

Conspiracy theories continue to circulate today in Pakistan, aided by the growing presence of smartphones and social media, leading to polio workers being regularly attacked – even killed – while out administering the vaccine. In 2019, false rumours spread about two boys fainting and vomiting after receiving a polio vaccine, leading to announcements over loudspeakers at mosques, angry protestors burning down a health facility and hundreds of children being rushed to hospital with complaints of fainting and vomiting supposedly following their polio vaccinations.

Somebody had filmed a few boys fainting and sent that around saying these vaccines are poisonous, says Dr Heidi Larson, Director of the Vaccine Confidence Project at the London School of Hygiene and Tropical Medicine. This vindicated beliefs held by some about the vaccine being a plot against Muslims by the West. 'People started running away from polio vaccination and dragging their kids out of hospital,' says Larson. The country saw a surge in new cases – more than 76 in 2019, when the figure for 2018 had been just twelve. Meanwhile, polio continues to occur in Afghanistan, largely in areas controlled by the Taliban, where the Red Cross and World Health Organization have been banned from operating. The unmonitored movement of people between Pakistan and Afghanistan also brings cases – and local rumours – into the country.

Despite these challenges, those leading eradication efforts have not given up and numbers have fallen by 99 per cent since their efforts began in 1988. In 2019, the world was declared to be free of two strains of the polio virus, hailed by all as a historic milestone. There are three distinct, immunologically unique strains of naturally occurring (or wild) poliovirus: wild poliovirus type 1, wild poliovirus type 2 and wild poliovirus type 3. They all cause the same symptoms, including irreversible paralysis and sometimes death, but their unique structures mean each type has to be eradicated individually to truly make people immune and rid the world of the disease. Type 2 was certified as having been wiped out in 2015 and type 3 in 2019. Wild poliovirus type 1, still circulating in Pakistan and Afghanistan (no new cases have been detected on the African continent since 2016), is the final hurdle.

The WHO's 'polio-free' designation relates only to wild forms of the disease, but to truly eradicate polio from the planet, another complex factor relating to the disease needs to be tackled – the fact that cases of polio can sometimes result from the vaccine itself. In some instances, someone who has received the vaccine can spread the virus to others who have not been vaccinated, known as vaccine-derived polio virus. This happens only on extremely rare occasions and it only spreads in under-immunized populations, with health experts highlighting this as a further need to ensure entire populations get immunized. Here, polio cases result from vaccination because the oral polio vaccine given to people contains live, weakened strains of poliovirus. The virus replicates in the gut and enters the bloodstream, eliciting an immune response to create protective antibodies in both bodily regions. This provides protection to the person being vaccinated (typically a child) but also to others around them as the gut antibodies limit the amount of time the virus replicates in the gut, reducing the amount of it passed out in their faeces, which could infect others who come into contact with it.

Even so, a child still excretes the virus for six to eight weeks after vaccination, just as they would if they had contracted the disease itself, albeit in a weakened form. In some cases this could provide passive immunity to those who come in contact with it, but receiving the vaccine itself is a much stronger means of protection. However, if the vaccine virus circulates in this way for prolonged periods of time, usually for more than one year, the virus can evolve and mutate to become virulent and able to cause disease. This is known as circulating vaccine-derived polio virus. It is rare and, again, only occurs in populations with low levels of immunity – due to low vaccine coverage – and more so in dense populations with poor sanitation as this enables the vaccine virus to spread continuously. A fully immunized population is protected against both vaccine-derived and wild polioviruses, according to the WHO.

In 2016 the vaccine used by the eradication programme was switched to one containing just wild poliovirus types 1 and 3, as type 2 had been wiped out. This resulted in significantly fewer vaccine-derived polio cases as type 2 was believed to cause almost 90 per cent of these. But cases from vaccines continue to occur, including the circulation of type 2 strains, which now spread more rapidly given that children are no longer being given vaccines protecting them against that type. The Democratic Republic of Congo saw a relatively large outbreak in 2019 with more than 35 cases; Nigeria had more than sixteen cases and many other countries across Africa reported cases that year. In any situation, irrespective of the strain, circulating vaccine-derived polio is a concern and must be handled in the same way as any polio outbreak, with immunization campaigns. In cases where type 2 is circulating, supplementary activities are needed, delivering vaccines containing just wild polio type 2.

Given the risk of vaccine-derived outbreaks occurring, the goal is to stop use of the oral polio vaccine once all three strains of wild poliovirus have been eradicated, instead using an inactivated form of the vaccine (IPV) that contains killed forms of the virus, which cannot mutate to become virulent. The obvious question is why these are not used already; the reason being that this form of the vaccine is injected – and therefore requires more skilled staff to administer – and more expensive, but, importantly, it only protects the people receiving it and does not prevent spread within a community, because virus replication is not limited in the gut, only the blood, meaning if someone vaccinated with IPV is infected with wild poliovirus, they won't develop the disease but they will still shed it in their faeces, whence it can then be transmitted to infect others. Many industrialized countries are already using IPV because the tiny risk of paralytic polio associated with the use of the oral vaccine is nevertheless deemed greater than the risk of an imported wild virus, for which rates are scarce throughout much of the world.

Despite the availability of an effective vaccine, complex biological and societal challenges have impeded efforts to wipe out the disease, pretty much entirely due to low vaccination coverage, and are now causing outbreaks just as final efforts seemed to be making progress. In an effort to support the final road to eradication, polio was declared to be a public health emergency of international concern (PHEIC) by the WHO in May 2014 due to the risk of international spread, and has continued to be considered as such by the emergency committee advising the organization. Some experts consider this decision controversial, explains Heymann, because PHEICs were never envisioned to be a chronic long-term recommendation.

In 2019, the WHO said the committee was gravely concerned by the significant further increase in polio cases globally in 2019 and that 'multiple circulating vaccine-derived polio outbreaks on the continent of Africa are as concerning as the wild polio situation in Asia'. The international threat remains and the PHEIC declaration stays in place, because today's teams need all the help they can get to rid the world of polio once and for all.

The last worm

The world had looked set to be free of the gruesome and debilitating disease Guinea worm, until an influx of last-minute setbacks emerged, meaning eradication would not be reached by the target date of 2020.

Guinea worm, also known as dracunculiasis, is a painful scourge where infection with parasitic larvae leads to metre-long worms bursting through the skin to lay their eggs. 'I marvel at how people stay sane with this disease,' says Dr Donald Hopkins, the Carter Center's special adviser for Guinea worm eradication and a key architect of the international eradication campaign for the disease. (The Carter Center, founded by former US President Jimmy Carter, works on a number of health and peacekeeping programmes worldwide.) 'I really think I would lose it mentally if I had to live with even one worm hanging out of my body, knowing I might also have another worm or several more worms to come out.' The record to date is held by a Nigerian man who had 84 worms emerge in one year, explains Hopkins, adding that in south-eastern Nigeria, communities once referred to the disease as the 'silent magistrate', because people dreaded waiting to see what sentence would be handed out to them by the disease that year. To further highlight the point, *dracunculiasis* is Latin for 'affliction with little dragons'.

The origins of Guinea worm stem far back in history, to biblical and Egyptian times. Ancient Egyptian medical texts show evidence of the worm's existence and the 'fiery serpents' sent to attack the Israelites in the Bible are considered to have been Guinea worms. The worms have also been found in mummies dating back to approximately 1000 BC. Now, more than 3,000 years later, they are on the brink of extinction , as efforts to eradicate the disease have been very successful.

In 2018, there were just 28 cases across three countries – Chad, Angola, and South Sudan. The Angolan cases are of particular interest because they were the first ever cases to be reported in the country, confusing experts as Angola is located far from any countries endemic for the disease. Numbers increased in 2019, however, with 53 human cases reported across the same three countries as well as Cameroon, with the overwhelming majority in Chad, which reported 47 cases.

But the case numbers today are still remarkable considering that in 1986 there were 3.5 million cases across at least 20 countries in Asia and, predominantly, Africa. If attempts are successful it will be the first parasitic disease to have been eradicated and, perhaps more importantly, the first disease ended without a vaccine or a drug.

Guinea worm is spread through contaminated water supplies – typically stagnant water such as ponds or drying riverbeds – that people living nearby have no choice but to consume. When a worm emerges from a person, usually on their lower leg, it does so through a painful blister that people often try to soothe by placing their leg in water. Once immersed in water, the worms release thousands of larvae, which are in turn consumed by water fleas living in the water. The larvae mature once inside the fleas to become forms that are ready to infect humans. When people then drink this water, the fleas are killed in their stomach, releasing the larvae, which develop inside the body into adult worms. It's the female worms that migrate down into the legs to burst out and lay their eggs, continuing the cycle of transmission. People do not become immune once infected, causing many people to suffer repeated infections, and there are no drugs to kill the larvae or worms inside the body. The only means of tackling the disease is through prevention: wound treatment; worm removal and bandaging; educating people to avoid drinking contaminated water supplies, but also to avoid contaminating water supplies with their wounds, training people to filter water and remove the water fleas, using chemicals to kill the larvae in the water and, ideally, improving overall drinking water supplies.

The fact that people drink stagnant water gives a window into the types of community affected by Guinea worm disease today: the poorest, most remote and inaccessible communities in rural Africa. This makes the final road to eradication the most challenging, explains Hopkins, who also worked on the smallpox eradication programme. These countries are last 'because they're the toughest ones', he says, giving reasons such as regional insecurity preventing teams from being able to access communities. The resulting migration fuelled by conflict also aids transmission. 'We're having to refine, adapt, modify and innovate strategies to deal with these new conditions,' says Hopkins. The added challenge, however, is the year-long incubation period of the disease. One missed case means waiting another year to see how many people become infected by that case and how far those infections may have spread. If they spread far it will be another year or two before you find those locations and then the cases. 'That makes Guinea worm a lot more challenging than smallpox,' says Hopkins.

These challenges are why the target date for eradication has moved several times. The first goal was 1991, then 2009, then 2015, then 2020. Now the goal is 2030. Recent setbacks include cases suddenly appearing in new countries and causing confusion, such as Angola, while other countries have seen the disease return after several years of absence, as was the case in Chad. Mali has not seen cases for a few years, but insecurity means some parts of the country have not been reached to know for sure. Mali, Chad and Ethiopia, which previously reported multiple cases in humans, recently further surprised public health teams with infections found to be circulating in cats and dogs, predominantly dogs, revealing a whole new layer to the disease and its transmission. 'It is evident that we will not have interrupted transmission by 2020 globally,' said Dr Dieudonné Sankara, Team Lead of the WHO's global programme for dracunculiasis eradication. Based on the current evidence, Sankara has said that he expects to achieve the target by 2030, or even before.

The number of animals infected is also extremely variable and new animal hosts are emerging to add even more complexity. In 2018, Ethiopia reported sixteen infections in dogs and cats as well as one infection in a baboon, Mali saw twenty infections in dogs and cats and Chad saw an unbelievable

1,040 infections in dogs and 25 in cats, according to the Carter Centre. In 2019, more than 1,970 dogs were again infected in Chad. Dogs are now considered a reservoir for the disease, particularly in Chad, and transmission between them and from them into humans needs to be stopped.

As noted previously, Chad was the main country affected by Guinea worm in 2019, reporting 47 of the 53 cases found in humans as well as the excessive infections among dogs. Here, eradication teams are also dealing with a unique mode of transmission – not through people drinking contaminated water, but instead eating raw fish entrails and poorly cooked aquatic animals that contain parasitic larvae. Control has involved treating water with chemicals to kill larvae, burying leftover fish guts to prevent dogs eating them and tethering dogs to force their worms to emerge in order to remove them.

The revelation in cats and dogs was a curve ball that hit eradication teams just as they thought they might actually meet their end goal of 2020. One of the reasons the disease was first chosen as a target for eradication was the understanding there was no animal reservoir. That was 'a real setback for the programme', says David Heymann of the London School of Hygiene and Tropical Medicine. Heymann explains that it's now not clear what will happen, as animals have many of the same habits as humans, such as walking into water to drink or bathe. Any circulation of Guinea worm in the animals might impede eradication, he says. But Sankara believes it can be stopped, by targeting the common enemy to both animals and humans: stagnant water. Remove or treat this and you're winning. 'As far as Guinea worm disease is concerned, the specific phenomenon is happening in localized areas of few countries,' he says. So if we control these, we control transmission in dogs and then the disease overall.

4 NEW AND UNKNOWN

From the point of view of an infectious agent, be it a bacterium, virus or any other pathogen primed and ready to attack, nothing beats the element of surprise. Unleashed upon an unsuspecting population, almost any infection will thrive as people first spend time racing to identify it, then seek ways to stop it and only then implement the relevant control measures. Make it airborne and the results are likely to be catastrophic, leaving health agencies and institutions worldwide rushing to develop tools and strategies to somehow control an unpredictable future. This is the power a new, unknown infection holds.

The increasing risk of a novel pathogen causing a global pandemic is in large part due to three elements of modern-day life: globalization, population growth and climate change. Countries have become more connected than ever in trade, travel, commerce, and more, fuelling greater movement of people internationally. 'The emerging diseases themselves haven't changed a whole lot, but human behaviour has, human mobility has,' says Mike Ryan, Executive Director of the Health Emergencies Programme at the World

Health Organization. As the global population continues to grow, people are also living in denser spaces, often closer to animals and with poor healthcare and sanitation facilities. 'We're stuffing billions of people into poorly managed areas,' says Ryan. Alongside this, rising global temperatures are further aiding disease vectors, such as mosquitoes, to survive across seasons and in new locations. The elements for a large-scale outbreak, possibly a pandemic, are all there, waiting for a pathogen to arrive and complete the mix – and global health experts know this. For example, it's predicted that if a highly contagious, airborne pathogen resembling the 1918 Spanish flu were to emerge today, an estimated 33 million people would die worldwide within just six months.

However, living in a climate where people await this form of impending doom can equally lead to an 'epidemic of over-reaction', says Ryan, who explains that the fear and societal reaction to a new outbreak can often be more damaging than the epidemic itself. In 1997, we saw the first human case of Avian influenza (bird flu). A three-year-old boy died in Hong Kong from a strain of H5N1 Avian influenza, which had been detected in geese and chickens the year before. This was the first known case of the H5N1 virus jumping from birds into humans and it went on to infect several people in Hong Kong and around the world. There were over 250 cases and 150 deaths in at least 55 countries. The arrival of this cross-species infection was indeed something to worry about, as was the spread of cases globally, but the infection required prolonged contact with infected birds and did not pass easily between humans, making rapid spread unlikely. The world did not see a global pandemic of H5N1, but it was on edge waiting for one. Re-emergence of the strain in 2003 saw case numbers among animals rise between 2004 and 2006, as the virus spread to wild bird populations in Africa and Europe. Human cases did occur, with almost 250 cases over this period and people began to panic, buying the antiviral drug Tamiflu in preparation, sometimes from unauthorised sources. National and international authorities had to step in and urge people to stay calm.

From a public health perspective, the 1997 outbreak did have some positive outcomes as the shock of seeing H5N1 cross the species barrier led to better programmes and resources for influenza monitoring, surveillance and preparedness. Surveillance efforts showed that after the surge in 2004–06, case numbers stayed in the tens for many years, then after a sudden rise to 145 cases in 2015, plummeted to ten cases in 2016, zero cases in 2018 and one case in 2019. But there are many strains of avian influenza under surveillance, not just H5N1, and even more strains of influenza among other animals, all of which could pose a risk to human health.

'There is no doubt that diseases will emerge, mainly from the animal kingdom, and they will rip through human populations. That has happened enough times in the past,' says Ryan. Away from influenza, he says that an airborne version of Ebola or a form of SARS that's even slightly more transmissible would be enough to 'bring our society to a halt'. The nationwide lockdowns, border closures, school closures, cancellation of events and significant hits to businesses and the global economy following the emergence of COVID-19 have proved Ryan's prediction to be true.

But such events are rare in the grand scheme of things, Ryan points out, with his team spending most of their time managing ongoing emergencies such as cholera and the malaria epidemic. Most cases of unknown diseases also don't typically become an issue: they occur in rural areas, at the human–animal interface, they don't spread very far and are usually only identified anecdotally, Ryan explains. 'That's a normal scenario.'

But, he warns, each of these seemingly mild infections is like Russian roulette. 'How many times can you pull that trigger, without it being the big one?' he asks. The problem when a 'big one' does then arrive is the speed at which we can be ready to defeat it. We're the most mobile human population that's ever been on the planet and it takes us years to develop a vaccine, explains Ryan. 'We're going to struggle.'

In a bid to reduce the struggle, the Coalition for Epidemic Preparedness Innovations (CEPI) launched in 2017, a public–private alliance prioritizing the accelerated development of vaccines against emerging infections. The alliance states that 'vaccines are one of our most powerful tools in the fight to outsmart epidemics', citing the 2014–15 Ebola epidemic in West Africa as an example of when a readily available vaccine would have saved thousands of lives. A vaccine that had been in development for many years, called rVSV-ZEBOV, had not yet been adequately trialled as the market for it was thought to be limited, and testing it in the field was also a challenge. The vaccine was only trialled once the epidemic was underway, and soon proved to be 100 per cent effective. It was then introduced over a year into the epidemic, saving thousands of lives, but thousands of people had already died.

Learning from this, CEPI prioritized six diseases for which it supported vaccine development, all known to harbour the potential to cause serious epidemics yet with no effective countermeasures available. These are MERS-CoV, Nipah, Lassa fever, Rift Valley fever, chikungunya – all viruses typically found in developing countries – and 'Disease X', the latter representing an unknown, or novel pathogen. These were chosen from a larger list of eleven priority diseases named in the WHO's annual research and development blueprint in 2018 (which also included the Zika virus), again chosen based on their epidemic potential and urgent need for better control measures, such as an effective drug or vaccine. It's a list that pertains to the international community, with Asian and African populations likely to be most at risk, but there is always the risk of importation to the US and Europe, particularly with mosquito-borne diseases, says Jan Semenza, Head of the Scientific Assessment sector at the European Centre for Disease Prevention and Control whose teams closely monitor trends in disease transmission across the continent.

Semenza explains that the ability for mosquitoes in Europe to transmit certain diseases on the list, such as Zika, has improved. 'We see more tropical diseases in Europe, in part due to the climate and air passenger volume,' he says. In late 2019 the first native cases of Zika were reported in France, meaning people had contracted the infection from mosquitoes within the country, not by visiting an endemic region – the first time this had occurred in Europe. But because the region was going into winter, the risk of continued transmission was considered low, as the mosquitoes carrying the virus would inevitably die out for the season. But the potential for it to return remains. 'We live in a complex multifactorial world that contributes to these types of risks,' says Semenza.

Even within the WHO list, and the CEPI list, further prioritization is needed to give order to the development of new vaccines, treatments and diagnostics. At CEPI, three diseases receive the primary focus, the first being MERS-CoV, a virus that typically spreads from animals (most often camels) to humans, but can also transmit between humans. Infection causes a respiratory syndrome, with more than 2,400 cases worldwide since it was first reported in 2012, of which more than 840 died. The majority of cases are in Saudi Arabia, but the virus has been reported elsewhere, with a large outbreak in South Korea in 2015, showing the potential for international spread.

Nipah and Lassa fever are the other two diseases being focused on first, both of which are spread by animals – bats and rats – as well as between people and both of which have a significant burden today, in Asia and West Africa respectively. All three diseases have no treatments and are only controlled though prevention measures. Vaccines are now being prioritized, with several candidates at the research stage. However, 'you could question whether those are the right targets', says Dr David Heymann, Professor of Infectious Disease Epidemiology at the London School of Hygiene and Tropical Medicine. He highlights that none of these three chosen diseases spreads particularly easily between humans, meaning they 'probably won't ever cause a major outbreak'. They spread most easily from the animals that carry the viruses, meaning you may have to focus the vaccination efforts on communities that live alongside these animals, such as rats in the case of Lassa, or on the animals themselves, such as camels in the case of MERS.

The biggest challenge perhaps for the WHO and CEPI is a strategy that prepares them for something they have not seen yet – for Disease X. How do you prepare for the unknown? With a blanket weapon you hope can be tweaked to tackle anything that comes your way, in this case a platform technology where an effective vaccine base can be built upon using proteins from the virus in need of control, creating a new, targeted vaccine each time. This could enable a new vaccine to be developed within months rather than years. Efforts were soon underway for COVID-19 when it started to spread rapidly at the beginning of 2020. But Heymann points out that even if, for example, the backbone of a vaccine for dengue can have something similar, like Zika, slipped in to make a new vaccine, the fact that it will still take many months means the outbreak will have spread. Outbreak containment is still a priority, he explains.

National agencies have also started efforts to prepare for Disease X, including Public Health England, which in 2019 included novel pathogens and emerging infections in their strategy to fight current and future health threats. Over the last ten years, twelve diseases were reported in England for the first time, the health agency said, including swine flu and MERS-CoV, but also Crimean–Congo haemorrhagic fever, a virus spread by ticks or contact with infected animal tissue, usually seen in Asia, the Middle East, Africa and the Balkans. This haemorrhagic disease causes an extensive range of symptoms including vomiting, diarrhoea, bleeding into the skin and an enlarged liver and is also on the WHO priority list. The UK also saw a large number of cases of the novel coronavirus and a comprehensive action plan was released in early March 2020 to help contain the outbreak.

Semenza is equally worried about the potential for imported known diseases like Crimean–Congo haemorrhagic fever and a completely novel pathogen that strikes having perhaps crossed a species barrier elsewhere, like COVID-19. In Europe his team sees extensive importation of tropical diseases, the majority of which – as much as 61 per cent – is driven by travel and tourism, with tourists coming into Europe and Europeans visiting parts of the world endemic for infections. Some diseases are further driven by changes in climate.

The very nature of emerging infections, however, is what makes them challenging. How can experts ever truly stop something they don't see coming and then can't understand quickly due to their complexity? Therefore, as well as vaccines, the attention now given to these diseases means they also have support for increased research to study the viruses themselves and how they are spreading, the development of diagnostics to confirm cases sooner and better surveillance to identify outbreaks earlier.

Semenza believes the solutions lie in facing our reality head-on and managing the problem upstream, preventing outbreaks from happening in the first place. 'Globalisation is here to stay; there's no point trying to stop air passenger volume,' he says, adding that we simply need to stop it being a risk factor. For example, with vector-borne diseases this can happen by identifying the high-risk months for disease importation – for Europe this is August to September when mosquito populations are at their peak – and identifying the regions seeing the most passenger arrivals and departures. Countries must then increase active surveillance systems at airports in these high-risk locations during this period. The public are also informed on how to spot symptoms for the diseases in question, such as dengue. 'Once someone is picked up who is symptomatic, you run tests and if they are positive you make sure they are not bitten by a mosquito and you treat them accordingly to stay ahead of an epidemic,' says Semenza. You then expand this for all vector-borne diseases, preventing outbreaks by taking people out of the epidemic.

At the end of the day, however, as well as the select list of eleven disease pathogens chosen for their epidemic potential, thousands of other pathogens exist and the final one on the list, Disease X, could be anything, something

Semenza describes as a 'black swan event': something rare and improbable with extreme impact, but 'very difficult to predict and prepare for'.

It's likely one of them will strike, and regularly, so officials need to learn from the past and be ready for them all. The good news, says Semenza, is that if we improve our public health preparedness activities and our core capacity for surveillance and response, we are able to lower epidemic risk. He conducted studies that found that a 10 per cent increase in core capacity could lead to a 19 per cent reduction in infectious disease threat events in Europe. The need now, then, is to boost that capacity.

Facing the unknown

It should be clear by now that disease outbreaks are anything but rare. In any given month, the World Health Organization reports news of anything from two to twenty or more outbreaks occurring at various locations around the world. In May 2019, for example, news was shared on fifteen outbreaks around the world, ranging from polio in Iran and Rift Valley fever in France to measles in multiple locations and monkeypox in Singapore. On one day, 9 May, three different outbreaks were happening: Ebola in the Democratic Republic of Congo (an ongoing outbreak of international concern), measles in Tunisia and MERS-CoV in Saudi Arabia. In some situations, cases are spotted easily, for example during an ongoing outbreak like Ebola or when it's a well-known infection for a region, such as MERS-CoV. The ever-growing presence of measles globally also has many clinicians on high alert, ready to identify the disease. But when a disease like monkeypox, which is native to Africa and is usually contracted through contact with infected animals, presents itself in Singapore, or when chikungunya, a debilitating infection of the joints most often found in Asia and the Americas, is diagnosed in Italy, health teams need to act fast. Even a disease known to a country can cause confusion when it arrives in a new setting, such as the capital city.

Here are some recent examples of established diseases in one region that have taken health officials in other regions by surprise.

May 2019: First imported case of monkeypox in Singapore

At the end of April, a 38-year-old Nigerian man came to Singapore to attend a workshop on the final two days of the month. On the second day, 30 April, he developed a fever, chills and skin rash, causing him to lock himself away in his hotel room for the week that followed. But his condition worsened, and he was taken to hospital by an ambulance on 7 May and put in isolation at the National Center for Infectious Diseases. Healthcare workers treating the man had to wear personal protective equipment.

To find out the cause of his sickness, laboratory tests were conducted on samples from the man's skin lesions, revealing that he had monkeypox, a rare viral infection described as a milder version of smallpox. The disease is native to central and West Africa and typically occurs through contact with the blood and bodily fluids of infected animals. Before travelling to Singapore, the man had been to a wedding in southern Nigeria, where he may have eaten bush meat harbouring the virus, Singapore's Ministry of Health said at the time. Thankfully, transmission of monkeypox between people is limited, making it unlikely the infection would spread enough to cause an outbreak. But the risk is still there, meaning anyone who'd had contact with the man had to be traced and, if considered to be a close contact, offered a vaccine and quarantined for 21 days. There is no vaccine against monkeypox, but the smallpox vaccine is considered to be effective in either preventing the disease or minimising the symptoms.

Investigations by the Ministry of Health identified 23 people as close contacts: eighteen who either attended or trained at the workshop, one person who worked at the workshop venue and four staff at the hotel. One person who attended the workshop had left the country and returned to Nigeria before investigations had begun, so they had to be followed up in their home country. During an outbreak, standard procedures also include surveillance measures and communication with the public. With all this in place promptly in Singapore, along with the quarantine of the case and his contacts, the risk of a monkeypox outbreak was deemed to be low by both the Ministry of Health and the World Health Organization. The virus had been stopped.

May 2017: Unexplained cluster of deaths spark fear of Ebola

On 25 April 2017, the WHO was informed of a cluster of deaths in Sinoe County, Liberia. Their cause was a mystery. An eleven year old had been admitted to hospital on 23 April with vomiting, diarrhoea and mental confusion after attending a funeral the day before, and had died within one hour of arriving. By 7 May, there were 31 cases experiencing similar symptoms, of which thirteen died; all the cases knew each other somehow, either as friends or family, and all but two of the cases had attended the same funeral as the child.

The outbreak unsurprisingly ignited fear in the community as the country was still recovering from the large-scale Ebola outbreak that had struck West Africa in 2014, in which Liberia was worst hit, with more than 4,000 deaths and 10,000 cases. Funerals had been a common cause of infection during the Ebola outbreak because bodies remain infectious even after death. Now here was a cluster of cases displaying similar symptoms to Ebola, all having attended a local funeral.

The Ebola epidemic had at least left Liberia experienced at responding to infectious outbreaks, with emergency management and laboratory infrastructures now in place. This enabled officials to rule out Ebola as a cause within 24 hours of the first alert and to send blood, urine and plasma samples to the US Centers for Disease Control and Prevention for testing. The tests revealed the infectious agent to be *Neisseria meningitidis* type C, a bacterium that causes meningitis as well as other forms of meningococcal disease, common symptoms being a stiff neck, high fever, sensitivity to light, confusion, headaches and vomiting. Transmission is airborne through inhaling droplets people cough or sneeze into the air around them, hence the cluster of cases. But the ruling of meningitis came as a surprise to some experts. Since 2010, the MenAfriVac vaccine had been rolled out across the 'meningitis belt' of Africa, which includes Liberia, and cases had fallen rapidly ever since. But the vaccine only protects against type A bacteria, which was the most common form seen before 2010, giving type C the opportunity to surface in communities like Sinoe county, in turn taking them by surprise.

Summer 2007: Chikungunya sets up residence in Europe

Chikungunya is a relatively little-known disease, or at least it was before it got a foothold in Europe in 2007. The word *chikungunya* means 'to become contorted' in the Kimakonde language spoken by groups in south-east Tanzania and northern Mozambique; the naming of the disease stems from the stooped appearance people can acquire due to the joint pain caused by the infection. Other symptoms include nausea, fatigue, muscle pain and a rash. It's a viral disease found most often in Africa, the Americas and Asia and is spread through the bite of the *Aedes albopictus* and *Aedes aegypti* mosquitoes, with the former being present in southern parts of Europe since early this century. The lack of a vaccine or an effective treatment make prevention the main control strategy available.

Imported cases of chikungunya had occurred in Europe for some time, brought in by travellers to and from endemic regions. All that was needed was for enough mosquitoes in Europe to pick up the virus that they would then transmit it natively, which is what happened in the northern Italian region of Emilia-Romagna in summer 2007. 'An individual who came back from India was bitten by a local mosquito, that local mosquito bit someone else and you had a big outbreak,' says Dr Jan Semenza. Between July and September that year more than 200 cases of chikungunya were reported. Native infections, known as autochthonous infections, have since occurred in France in 2010 and 2014, as well as in both France and Italy in 2017, with Italy experiencing a second large-scale outbreak, seeing more than 300 cases across three areas – Rome and Anzio in the middle of the country and Guardevalle in the south. 'This is air passenger volume that drives that kind of importation risk from endemic areas into Europe,' says Semenza. 'Then people within Europe travel and move around and so you have local spread.'

2007 was a first for both Italy and Europe and though public health experts had long feared mosquito-borne diseases outstaying their welcome, they had to act fast when it happened, as the consequences of responding too late are vast. The virus circulating in the blood of the community 'could have threats to the blood supply,' says Semenza, as it could infiltrate the supplies available at blood banks. Authorities controlled the mosquito populations promptly, by blocking breeding sites and administering insecticides. They also educated the public on preventing bites and put a pause on blood donations – though this measure in itself has life-threatening consequences. Constant surveillance would also be required moving forward, particularly at the community level, as the communities affected by outbreaks commonly play a key role in identifying them. 'It's the communities that pick up on these outbreaks,' says Semenza. In the 2017 chikungunya outbreak, someone in the local community recognized the symptoms, having encountered them while visiting Asia, and brought it to the attention of health officials, Semenza explains.

For now, the temperate climate of Europe has helped keep mosquito-borne diseases like chikungunya in check, stopping them becoming truly endemic, because the mosquitoes die out once winter hits, breaking transmission of the virus between humans. But warming climates are helping extend the summer season and the global spread of chikungunya is making the chances of holidaymakers importing the virus even greater. Mosquitoes are spreading their wings, and everyone needs to be at the ready to spot them and stop them.

MOSQUITO DOMINATION 5

In a game of word associations linked to infectious disease, the word 'infection' might commonly lead to naming a disease that's airborne, like a cold or the flu, while the word 'mosquito' might often lead to the word malaria. This is a presumption, but highlights the point that when it comes to infections, certain diseases inevitably come to mind first, and they are most likely the diseases people are most affected by or most familiar with. But while eyes are on the attention-seekers, the quieter, chronic infections are given the opportunity to wreak havoc and take over the world while no one is watching – which is exactly what they've done.

In the last 30 years, the number of dengue infections – a viral disease causing fever and joint pain – has risen thirty-fold, with an estimated 390 million infections each year, and more and more countries reporting their first outbreaks. Half of the world's population is now thought to be at risk of the infection – we'll look more closely at dengue later in this chapter. Another example is West Nile virus, a virus spread by a combination of migrating birds and mosquitoes (the insects biting both the birds and humans), which, before 1999, few Westerners had heard of despite it being present in many regions, including Africa, East Asia and the Middle East. The virus was imported into New York in 1999 and soon spread rapidly, establishing itself across the United States, into Canada, and down as far as Venezuela in subsequent years. The US has seen more than 50,000 cases of the fever and body-ache-causing virus since it was first imported, and despite most people usually showing no symptoms, almost half of the cases in the US developed a severe form of the disease, where symptoms include high fever, disorientation, coma, convulsions and paralysis; more than 2,300 died. First reports of West Nile virus in Europe occurred in 1996, with cases then reported regularly since 1999. In 2018, there was a large outbreak, causing more than 1,600 cases across sixteen countries, compared to 300 cases across three countries in 2017. 'That occurred unexpectedly, we had never seen such a big outbreak,' says Jan Semenza of the European Centre for Disease Prevention and Control.

Aside from sudden outbreaks, global takeover by these diseases has been slow and steady over many years; they have set up residence in multiple locations with ease, courtesy of one key element: the mosquito. More specifically, the *Aedes* mosquito, also known as the 'tiger mosquito' due to its white stripes. This genus of mosquito is a versatile disease vector now found on all continents except Antarctica, thanks to global trade, travel, migration, dense urban populations and warming climates. The females bite humans to obtain blood to produce their eggs (this is true of all mosquito-borne diseases) and in doing so harbour the ability to spread the two diseases named above as well as Zika, Rift Valley fever, yellow fever and lymphatic filariasis, also known as elephantiasis. Zika transmission has been reported in more than 85 countries worldwide, including the United States and more recently France. When it comes to this family of vector-borne diseases, spread by *Aedes* mosquitoes, Semenza 'is concerned', he says. 'They have epidemic potential and could have huge impacts on the population.'

Efforts to control the diseases focus on controlling the numbers of mosquitoes – for example, removing breeding sites, and use of insecticides – as most of the diseases do not have a treatment or vaccine, the latter being something research teams are working towards. In 2019, the World Health Organization introduced a strategy using sterilized mosquitoes in hope of further controlling mosquito populations. The 'sterile insect technique' involves generating large numbers of sterile male mosquitoes using radiation in a lab, which are then released in the wild to mate with females. Mating won't result in viable eggs being produced, meaning mosquito numbers then decline over time. The technique was already established as a form of pest control in agriculture, with favourable results. Variations on the approach, using bacteria to make the mosquitoes sterile, or genetically modified mosquitoes, have been trialled in certain settings, such as the United States and China, but with mixed results. The *Aedes* mosquito is the toughest of them all and is considered to be one of the greatest threats to public health today. Only time will tell if this combination of efforts will prove effective.

Four-faced dengue

As noted previously, more than half the world's population are thought to be at risk of contracting dengue, a virus spread by the bite of an infected mosquito that causes anything from severe headaches, joint pain and swollen glands to severe bleeding, persistent vomiting and rapid breathing. Some people show no symptoms at all, helping the disease to spread, while in rare circumstances the virus can prove fatal, making every infection a game of chance. Today, thanks to globalization and overpopulation, an estimated 3.9 billion people have no choice but to take the gamble. Because some people see no symptoms, the true number of cases is impossible to find out, but more than 1.7 million were reported to the WHO in 2019, though models estimate there could be as many as 390 million each year. It is the only mosquito-borne disease to have increased exponentially in recent years.

But dengue's takeover has been a gradual one, a tortoise in the race where airborne outbreaks were the hares. Before 1970, nine countries had experienced severe dengue epidemics and now more than 120 are fighting the disease, aided by warming climates, rapid and unplanned urbanization, and poor health services in these urban environments. Semenza believes the movement of people is responsible for the majority of epidemic events, but the movement of people also needs the movement, or existence of mosquitoes, as without them the virus cannot spread between humans. One type of *Aedes* mosquito, *Aedes aegypti* has long had a presence across the tropics and many parts of the subtropics, including the United States and Middle East, and its numbers are thriving due to the rise in dense urban populations – giving them plenty of people to bite – with poor water management that then leaves shallow water to help them breed and lay their eggs. A second type, *Aedes albopictus*, has undergone a 'massive global expansion' according to the European Centre for Disease Prevention and Control, and is now widely found in many new regions, including most of Europe, due to the global trade in tyres and in ornamental bamboo plants, both of which can provide shallow pools of water for mosquitoes to breed. *Aedes albopictus* is now listed as one of the top 100 invasive species. It spreads dengue less efficiently than its *aegypti* cousin but can spread it nonetheless.

This flow of people and mosquitoes has inevitably helped the dengue virus spread. The Americas and Asia continue to be the most affected regions by far, experiencing millions of cases each year, and with Asia facing 70 per cent of the dengue burden. But cases are now reported on every continent except Antarctica, both imported and locally transmitted. Europe saw a large-scale outbreak on the island of Madeira, Portugal in 2012, after the virus was imported onto the island, resulting in more than 1,000 cases as well as 78 in other countries among people who had visited Madeira. Europe as a whole has seen more than 2,000 cases each year since 2015, the large majority of which are imported cases. For example, in 2018 a total of fourteen locally transmitted cases were reported in France and Spain, but this shows the mosquito vector is present and capable of causing greater damage when the conditions are right. The situation is similar in North America. The US reported more than 860 cases in 2019, with at least twelve of these transmitted locally. With its ability to cause both no symptoms and severe disease, Dengue has slipped in and caused just enough harm to establish itself as a force to be reckoned with – but the virus also has another weapon within its biology, ready to catch you off guard.

Four types of dengue virus, called serotypes, are known to cause dengue infections; these are named DEN-1, DEN-2, DEN-3 and DEN-4. They share 65 per cent of their genetics, but are just different enough that each one can infect and cause symptoms, even if you've already experienced one of the others. Infection with one type, let's say DEN-1, leaves you immune to future infections with DEN-1 but not to the remaining three serotypes. While the different types were once separated and found in different geographical regions, all four now circulate together in a growing number of tropical and subtropical locations. Someone may therefore believe they're protected having survived a previous infection, when there's in fact a three in four chance they are not.

To make matters worse, experts hypothesize that the residual immunity after a first infection leaves you more prone to developing the severe form of the disease, which can cause extensive bleeding or organ impairment, should you become infected with a second subtype – a theory called antibody-dependent enhancement. A 2017 study by US and Nicaraguan researchers added further evidence to support this theory, finding that when antibody levels from a former infection fall to an intermediate level – not a low level – they increase the chances of severe dengue. The researchers suggest that perhaps the intermediate amounts of antibody aren't enough to stop the virus, but may instead bind to them and help them reach susceptible cells. Either way, the findings supported the idea that prior infection puts you at greater risk of severe disease second time around. Thankfully, infection with a third subtype may not be severe, and is in most cases mild, says Dr Raman Velayudhan, who works on Vector and Ecology Management at the WHO.

The WHO has set targets to reduce illness from dengue by 25 per cent and mortality by 50 per cent by 2020, the main strategy for which involves the control of mosquito populations and educating the public on such control as well as bite prevention. The release of sterile male mosquitoes to mate with females who in turn lay eggs that don't hatch (as discussed earlier in this chapter) is also currently being trialled. But a vaccine is once again the ideal.

A deeper understanding of the complex immunology underlying the disease is hoped to further the development of an effective vaccine against dengue, as the disease currently has no treatment. 'The presence of four serotypes of dengue virus poses a great challenge to get a perfect vaccine,' says Velayudhan. There is one vaccine now licensed for use in at least nineteen countries, Dengvaxia, developed by pharmaceutical giant Sanofi Pasteur. Approved for use in people aged nine to 45, the vaccine showed varying protection levels against the four different serotypes of the dengue virus, but acceptable levels of protection overall, showing an efficacy of 59.2 per cent against all serotypes of the virus in its phase III trials, and 79 per cent protection against developing the severe form of the disease. Countries with high burdens of the disease were excited and promptly approved the vaccine for use, but some countries too promptly, as Dengvaxia was soon shrouded in controversy after a disastrous, fatal, introduction in the Philippines.

In 2017, the healthcare authorities of the Philippines brought Dengvaxia into their nationwide routine vaccination schedule to protect the school-age population from the debilitating disease. The country had reported more than 175,000 cases of dengue the year before. Once the campaigns were underway, reaching more than 800,000 children, Sanofi Pasteur released findings from a new analysis of their vaccine, showing that it put people who had never been infected with dengue before at a greater risk of developing the severe form of the disease – again linked to the unusual immunology of the disease. The vaccine essentially acts like a first dengue infection, meaning it only provides protection to people who have already been infected once. This makes the timing of administering it crucial to get right and really requires the development of a rapid diagnostic test, which does not yet exist, to deliver it appropriately in endemic countries. Some experts consider this to be a serious setback for the vaccine, but there is still a benefit at the population level as countries with such high burdens of the disease are likely to have a large proportion of their population already infected once.

The finding was nonetheless worrying to Filipino populations and the vaccine was linked to the deaths of several children and soon banned in the Philippines. The controversy led to criminal investigations, congressional hearings and a surge in overall vaccine hesitancy among parents, not just against Dengvaxia but also vaccine-preventable diseases such as measles. Myths around vaccines have since flourished. Thousands of children reported adverse effects following vaccination and the Philippines department of health was forced to provide the budget for their treatment. In 2019 the country saw dengue surge even more strongly than usual, with more than 320,000 cases and over 1,200 deaths, forcing authorities to declare a national epidemic and contemplate using the vaccine again, but it wasn't brought back in.

Dengvaxia remains licensed in all the other countries that approved its use, with the United States joining the list in 2019, but with clear guidelines for use solely in people who have been infected with the virus before. The European Medicines agency and US Federal Drug Administration (FDA) further stress this in their guidelines for use, but the FDA also only licensed the vaccine for use in people aged nine to sixteen living in endemic areas with previous infections confirmed by laboratory tests, after a large-scale trial found Dengvaxia to be 76 per cent effective in this age group. Overall, the caveats of the vaccine are transparent and experts, including the World Health Organization, support its use in controlling the growing epidemic. In 2019, South Asia faced extreme levels of the disease, with Bangladesh reporting its worst outbreak since it first began reporting cases, experiencing more than 92,000 infections. The situation is far from under control.

Last year, Dr Soumya Swaminathan, WHO Chief Scientist, said that despite the organization's best efforts, current efforts to control dengue are falling short. In 2019, more than 3.7 million cases were reported to the WHO – 2 million more than 2018. Vector control is not enough, with methods often being unsustainable or applied incorrectly on the ground. 'We desperately need new approaches,' he said, advocating for the use of sterile mosquitoes, as well as more effective vaccines, diagnostics, mosquito control, treatments, all of it in an upgraded form to stay not just one, but several steps ahead of the virus and its vector. The only way to stop this widespread disease is to attack from all angles.

Zika, Zika, everywhere

Dengue may have shown its infectious peers how to sneak in beneath the radar and dominate the world in one long, gradual takeover, but another virus in the same arbovirus group recently showed that it's possible to cover the same ground in one fell swoop. After decades in dormancy, the Zika virus became transcontinental in a matter of years and then global in a mere few months in 2015.

Identified in Uganda more than 70 years ago in 1947, the Zika virus was first discovered in monkeys, then in humans five years later – again in Uganda, but also Tanzania. In the decades that followed, sporadic cases occurred in Africa and Asia, but the virus wasn't anything to worry about. Not yet at least. The first real outbreaks were reported in 2007 on the Pacific island of Yap, infecting an estimated 5,000 people, then in 2013 on four other Pacific islands in French Polynesia, this time causing an estimated 30,000 infections. But only in 2015, when the virus reached Brazil, almost certainly via a traveller from French Polynesia, and caused a national emergency, did Zika truly come to the world's attention. It then promptly took over the globe.

Zika's spread was at first aided by a strong choice of disease vector; the same one as dengue, chikungunya and yellow fever: *Aedes aegypti*, which thrives in dense populations or environments with stagnant or shallow pools of water, including areas with poor sanitation facilities – where, again, people are likely to be plentiful. Female mosquitoes with the ability to spread the virus were therefore ready and waiting in multiple regions, ready to bite the many unsuspecting carriers of Zika, as the majority of people – around 80 per cent – do not develop symptoms.

The virus soon spread beyond the known boundaries of the mosquito vector, thanks to a readily susceptible and mobile population with no prior immunity, as well as one more weapon: the ability to spread sexually, without a mosquito, though this is less common. By February 2016, Zika had flourished and spread across the Americas, with more than 20 countries and territories reporting local transmission, and an outbreak also underway in Cabo Verde, West Africa since October 2015. Further spread of the infection was predicted to occur through international travel and the World Health Organization declared the situation to be a public health emergency of international concern (PHEIC). The global fight to end Zika then began.

In the beginning, Zika was shrouded with mystery and its declaration as a PHEIC made because of the many things that were not known about the virus. Its characteristics had changed over time, as it gained geographical momentum. 'Zika was not a pathogen that was considered that highly pandemic, and in 2016–17 it proved to be highly pandemic,' says Sylvie Briand, Director of the Infectious Hazard Management team at the World Health Organization. A small thing had the potential to become a bigger thing, and it did, she explains.

Symptoms of the disease itself were mild – fever, tiredness and an occasional rash – and most people didn't even develop them, so fatalities were rare. But the areas affected by the virus simultaneously saw a rise in the numbers of babies born with microcephaly – a smaller than normal head – and people affected by other neurological disorders such as Guillain-Barré syndrome, an autoimmune disorder affecting the nerves, eventually causing paralysis and sometimes death. In October 2015, for example, Brazil informed the WHO that 54 cases of microcephaly had been detected since August that year, in just three months, alarming experts and the public alike. This led to a retrospective investigation of the 2013 outbreak in French Polynesia, which identified seventeen newborns with a range of brain malformations. The public health emergency was therefore called on the clusters of microcephaly, not Zika, explains Heymann, who led the emergency committee of experts advising the WHO on the situation. 'It was not known whether Zika was associated, although it was associated geographically,' he says.

The virus had morphed from causing mild and infrequent disease to causing large outbreaks linked to neurological disorders. To stop it would require a much greater understanding of it. The need for prevention measures, such as mosquito control and diagnostic tests, was also stressed, particularly among pregnant women and women of child-bearing age.

The suspected links were soon confirmed. Reviews of the scientific literature and of reported cases firmly concluded that Zika virus during pregnancy is a cause of congenital brain abnormalities, such as microcephaly, as well as other complications, including pre-term birth and miscarriage, and the virus in general is a trigger for Guillain-Barré syndrome and other neurological complications. 'In a matter of a few weeks we were able to prove that yes, microcephaly was linked to Zika infection and we had already put in place some measures to control the disease spread,' explains Briand. The PHEIC was officially transferred over from focusing on microcephaly to the Zika virus itself after which it soon became clear that Zika was a global epidemic in need of a long-term control strategy. 'WHO set up its own Zika programme,' says Heymann, as the virus 'was no longer an emergency, it was a disease they had to deal with in the long term'. And long term it was.

Brazil was by far the country hardest hit by the epidemic, reporting an estimated 95 per cent of the world's cases. Within Brazil, the majority of infections occurred in the country's north-eastern states. By the end of 2016, more than 1.6 million cases of Zika were reported in Brazil, of which around 41,000 were among pregnant women; by then almost 2,000 cases of microcephaly had also been confirmed. This time period was the peak for infections across the Americas and globally, with South America reporting more than 30,000 cases per week in early 2016. By the end of that year numbers had dropped significantly to a few thousand per week. The United States reported more than 5,000 cases in 2016 then a few hundred in 2017, though here the infections were mostly imported among travellers, with just a small proportion contracted from local mosquitoes. Similar trends were seen in Europe with a little over 2,000 travel-associated cases in 2016, dropping to around 200 cases in 2017 and eighteen cases by 2018. Europe also saw 22 sexually transmitted cases over this time, and in October 2019 reported its first case of transmission from local mosquitoes, in France.

It should be noted that infections have long occurred in Asia, but the situation there is unusual. Zika has been circulating there since the 1960s, proven through blood tests and genetic testing among both residents and people who have travelled to the region, but the extent of the epidemic is largely unknown. Surveillance and monitoring is limited, regarding both infections and any syndromes linked to them, the WHO stated in a 2019 epidemiological update, adding that 'improved surveillance and epidemiologic investigations are needed to better ascertain the incidence of ZIKV [Zika Virus] infection in the South-East Asia region and its impact on birth outcomes.'

New cases of Zika continue to occur worldwide today and the virus is still spreading to new countries, albeit all at a much smaller frequency. In 2019, the number of countries reporting evidence of local transmission of Zika reached 87, with Ethiopia being the latest addition to the list in 2018.

More than 60 other countries that have not yet reported Zika transmission have established populations of the *Aedes aegypti* mosquito, and the WHO warns that the risk of the virus spreading there remains, meaning all countries need to stay vigilant.

However, it could be said that the true focus of the Zika epidemic is no longer about new infections, which are being monitored, but instead the thousands of people and children now living with the neurological complications of their infection, known as congenital Zika syndrome. Unlike most outbreaks or epidemics, where a treatment, vaccine or even just containment measures eventually leave communities free of the disease, the burden of Zika is not resolved once case numbers come down and the spread of new infections is curbed. There are tens of thousands of people who will have experienced nerve damage and muscular paralysis just days after their infection, with some cases resulting in complete paralysis, or even death, while others go on to heal adequately, leaving minimal symptoms. Longer-term, there will now be a generation of children with congenital Zika syndrome, most of them born with small heads and various disabilities relating to this for whom their parents or carers will need to provide lifelong care. Children born with microcephaly suffer a range of developmental problems, including feeding challenges, impaired movement, visual and hearing abnormalities and seizures. Many children without a small head may still suffer symptoms, such as brain calcification or muscle stiffness, and to make matters worse, many of the potential outcomes from infection are still unknown, meaning families can't fully know what to expect.

Cases of Zika, and congenital syndromes stemming from them, are predominantly among the poorest, most vulnerable, communities of any country, particularly in Brazil where healthcare services, including ultrasound scanning, are limited in rural settings and many mothers don't find out their children have microcephaly until they are born. In a 2017 report by Anis, the Institute of Bioethics in Brazil, anthropologist Debora Diniz highlights the stories of various women in the north-eastern state of Alagoas, a small state that experienced a disproportionate number of cases of congenital Zika syndrome. Diniz highlights the various fates of the families affected by what was known locally as 'the little mosquito problem', ranging from local discrimination, abandonment of babies and a general lack of knowledge about what exactly was wrong with the children to the struggles of teenage mothers with no means to care for their 'special children'. Financial support and therapy sessions to provide visual and audio stimulation as well as help children move or communicate are provided, on paper, but access, as with any government aid, is variable. As these children develop and grow, their parents will learn alongside the researchers trying to understand the situation.

The burden of Zika is not going anywhere. New cases have dropped significantly, to a trickle compared to the original waterfall, but they still occur, and travel restrictions remain in place for pregnant women, for example. Among the continued cases there will still be reported cases of Guillain-Barré syndrome and babies born with congenital Zika syndrome. For this reason, combined with the continued lack of treatment or a vaccine for Zika infection, the virus is on the WHO research blueprint of priority diseases and candidate vaccines are in trials. But the only real protection today is prevention, namely mosquito control, a strategy that ultimately also reduces rates of other diseases spread by these insect vectors, such as dengue and chikungunya. The world did see dengue numbers fall shortly after Zika emerged, as aggressive mosquito control efforts began, but they rose sharply again in 2019, so what will truly stop the scourge of mosquitoes remains a mystery for now.

6 TIME FOR A COMEBACK

Modern medicine has come a long way, evolving greatly to improve the health of populations over time. In its first iteration, the goal of medicine was simply to treat and prevent disease, using whatever means and skills were available. Aside from injuries, this was most commonly the control of infections that surged though communities aided by dense living conditions and poor sanitation. But modern medicine has led to a greater understanding of the human body and the multitude of infections and ailments that weaken it, and to the development of drugs to treat those maladies, most notably the development of modern-day antibiotics, which are now under threat.

The plant origins of many antibiotics have long been used by various cultures to fend off disease, but only in the nineteenth and twentieth centuries did scientists begin to understand that microorganisms like bacteria were behind these diseases, soon realizing that certain chemicals could then kill these bacteria. According to the Microbiology Society, the first antibiotic in this sense was the chemical arsphenamine, discovered by the German physician Paul Ehrlich as a treatment for syphilis in 1909. But the real game changer was the discovery of penicillin by Alexander Fleming in 1928, which would go on to become the first mass-produced antibiotic. Nicknamed 'the wonder drug', penicillin was soon used widely across the globe and led the way for the discovery of many more classes of antibiotics, such as fluoroquinolones and tetracyclines, each of which was, importantly, used to treat a wide range of infections, not just one or two. Their blanket use meant these drugs formed the backbone of modern medicine, treating everything from respiratory and chest infections, such as tuberculosis, to meningitis, septicaemia, urinary tract infections and acne, to name just a few.

Today, hundreds of different antibiotics exist. They have helped swathes of infections become history, or at least removed from our radar, but the potency of these drugs is not what it once was. Many antibiotics are now at risk of becoming redundant, threatening a century of progress, as bacteria are evolving to develop resistance against them, becoming 'superbugs'. Pathogens we thought we had beaten, or had under control, are returning with more force and human behaviours are in turn helping them flourish. Antimicrobial resistance 'is one of the few emerging public health issues which is known,' says Dr David Heymann, Professor of Infectious Disease Epidemiology at the London School of Hygiene and Tropical Medicine. 'It deserves much more attention than it's getting now.'

Let's take the example of tuberculosis, where drug resistance is a major public health problem. In chapter 3 we examined the current burden of tuberculosis and the challenges around controlling the disease today, but a key factor is the ever-growing issue of drug resistance. First came patients resistant to a crucial first-line drug, rifampicin, followed by people experiencing multi-drug-resistant tuberculosis, which encompasses additional resistance to another first-line and powerful drug, isoniazid. It's estimated there are around 500,000 cases of multi-drug-resistant TB each year. Then extensively-drug-resistant tuberculosis emerged, which is resistant to at least two more anti-TB drugs and people feared total drug resistance would soon be on the cards. People have indeed emerged with complete resistance. For example, there were four cases in India in 2012, a country that has the lion's share of TB cases, both regular and drug-resistant. Cases of total resistance were documented in Italy and Iran before that, but thankfully this remains extremely rare, even today.

While the development of resistant genes can occur naturally, human behaviours have helped microbes evolve faster. In the case of TB, as we have seen, curing oneself of the disease requires dedication, as treatment is a minimum of six months. Shorter regimens introduced in recent years have brought this down to four months in some cases, but originally, becoming free of TB would involve a daily pill for at least half a year of your life. Unsurprisingly, people started out keen and motivated but soon lapsed, possibly having felt better and believing the drugs were no longer necessary. Whatever the reason, many patients stopped taking their drugs earlier than the duration set out for them at the beginning, meaning the bacteria behind the disease were not fully wiped out. This allowed the remaining bacteria not only to replicate and grow in number, bringing the infection back, but also to mutate and develop resistance to the antibiotics they had already encountered.

The rapid replication rates of bacteria mean their turnover is high, and they are able to quickly evolve past a problem they encounter, such as a drug that is trying to kill them. The microbes that mutate and develop resistance to a drug will be the ones that survive and go on to proliferate the most. Bacteria are also able to transfer genes between themselves, including resistant genes that have emerged. Once bacteria are resistant, they leave a person untreatable by the drug being used, and the person is unlikely to get better. This means longer treatment, new or repurposed drugs where possible, and sometimes a longer hospital stay, all of which means more expense to treat those affected.

The situation with TB is representative of the situation across most bacterial infections, in that people not completing their course of antibiotic treatment has slowly rendered certain antibiotics useless. But a bigger contributor to the global problem we have today is the misuse, or overuse, of antibiotics by healthcare and agricultural professionals either to treat infections that didn't really need antibiotics, or to promote growth or prevent disease in already healthy farm animals. This widespread use of antibiotics has meant bacteria are exposed to a greater number and concentration of antibiotics, increasing their chances of developing resistance and thus thriving. It's a truly global concern, as the World Health Organization states that antibiotic resistance is present in every country. Heymann believes that solutions therefore also need to be global. 'The most important thing is to work together and find solutions that everyone can do,' he says. This includes reducing the overuse of these drugs, which can easily be bought over the counter or on the black market in many countries, as well as ensuring strategies are agreed upon by companies and animal authorities to prevent the misuse of antibiotics in agriculture. According to the WHO, approximately 80 per cent of the total consumption of antibiotics is in the animal sector in some countries, largely for growth promotion in healthy animals. A 2017 study in the *Lancet Planetary Health* found that interventions that restrict antibiotic use in food-producing animals reduced antibiotic-resistant bacteria in these animals by up to 39 per cent. But human prescriptions are also to blame, with one 2018 model by Public Health England finding that at least one in five prescriptions in England alone are inappropriate. The US Centers for Disease Control and Prevention estimates higher numbers for the United States, with almost one in three of all antibiotics prescribed to outpatients and in hospitals proving to be unnecessary.

Unfortunately the issue of resistance really stems across all pathogens – bacteria, parasites, viruses, fungi – hence the more common use of the term antimicrobial resistance today, instead of antibiotic resistance. But it's fair to say that resistance predominantly remains a concern among bacterial infections in terms of scale, though parasitic diseases, including malaria, and viruses such as HIV and influenza are seeing a growing amount of resistance against the drugs commonly used to treat them. Malaria has battled resistance to drugs since the first chloroquine treatments were developed in the 1940s.

The problem is real and it's already here, killing at least 700,000 people each year, one third of whom die from drug-resistant tuberculosis, a 2019 report found, adding that the figure could increase significantly to 10 million deaths per year by 2050. Beyond deaths, it will fuel financial crises and force tens of millions of people into extreme poverty. As with everything relating to disease, it's not just individual health at risk, but global health and security. But away from the gloom, experts have outlined actions that can help, on the individual and societal level. Individuals can practise good hygiene – such as washing their hands regularly – to prevent the spread of infections, ensure they finish their antibiotic courses if prescribed and ensure they are only prescribed when actually necessary. They can also try to buy meat from farms that do not use antibiotics on their animals. Larger-scale solutions need commitment from world leaders and heads of healthcare and pharmaceutical organizations. A major review on antimicrobial resistance, which concluded in 2016, set out ten recommendations to tackle resistance on a global scale by reducing unnecessary use, with a particular focus on four items: a global public awareness campaign, the supply of new antibiotics through increased research and development, improved diagnostic tests to ensure that antimicrobials are used only when needed, and reducing the unnecessary use of antibiotics in agriculture.

Resistance: the frontline

It can be hard to relate to the concern around drug resistance if it hasn't affected you yet, or when large numbers are used to depict the situation, like 10 million deaths by 2050, or 500,000 people developing drug-resistant tuberculosis; the latter in particular may seem a distant concern if you live in a community where tuberculosis just isn't an issue. But antimicrobial resistance (AMR) is, unfortunately, everywhere; it's already affecting your community, not just those far away – you may just not know it yet. For example, around 35,000 people die each year in the United States from antibiotic-resistant infections. That number is similar for Europe at 33,000, with the UK representing 5,000 of those deaths.

If nothing is done to address resistance as soon as possible, experts warn that everyone will be at risk as we embark on a post-antibiotic era – although particularly those with weakened immune systems due to age, disease or chemotherapy – because the scenarios in which problems will arise are everyday ones: treating simple cuts and wounds that are at risk of infection, surgery, childbirth, hospital stays and even having sex. Some of these scenarios are explored in the following pages.

Surgical risk

Without effective antibiotics, the success of major surgeries will be compromised, the World Health Organization warns. This is because infections at surgical sites – where surgeons enter the body and where artificial devices are inserted – are fairly common, for example during cataract surgery, joint replacement surgery, breast implant and pacemaker procedures and organ removal. Surgical infections are currently prevented using antibiotic prophylaxis: the use of antibiotics before surgery to prevent an infection developing. The idea is to ensure that antibiotics are already in a patient's body when an operation begins, particularly in the tissue where surgeons will make their incisions, and during the surgery itself, ready to kill any bacteria that arise. But this practice has also been subject to misuse and overuse. For example, antibiotics are often also used after surgery, which the WHO and many experts deem unnecessary and inappropriate on most occasions, suggesting overuse in this way increases the risk of resistant bacteria emerging. The global agency says as much as 43 per cent of patients are inappropriately continued on antibiotics after surgery. Some surgical teams also administer antibiotics at inaccurate times – either too early or too late, when they will have less effect – and may not use the most appropriate class of antibiotics. All forms of misuse increase the chances of bacteria becoming exposed to ineffective antibiotics that they can then develop resistance to. An estimated 33 per cent of surgical patients get an infection post-surgery, according to WHO data, and of these, as many as 51 per cent show resistance to antibiotics.

Studies show that wound infections are the most common form of hospital-acquired infection among surgical patients, and if antibiotics become redundant in first preventing these infections but then also in treating those that occur as a result, the outcome is bleak. Very bleak.

Fears for labour

Giving birth, having children, is at the core of human existence but it has always brought the risk of maternal and child mortality if things don't go to plan. In 2017, about 295,000 women died during and following pregnancy and childbirth, a figure the WHO calls 'unacceptably high'. Narrowing in on infections around childbirth, the mortality figure among mothers is more than 30,000. Among newborns it's much higher, more than 400,000, as their immune systems are not yet developed and half of these deaths are estimated to be due to infections that didn't respond to available drugs. Resistance is already having a devastating impact.

The crisis is largely affecting low and middle-income countries at the moment, with 94 per cent of maternal deaths occurring in these countries, but high-income countries are not exempt. Maternal mortality has been rising in the United States in recent years, for example, from almost 700 deaths in 1990 to more than 1,000 deaths in 2015. These are not all due to infections of course, but infections and now resistance play a role, accounting for as many as 13 per cent of these deaths in the US. Among high-income countries as a whole, around 5 per cent of these deaths are estimated to be due to infections, according to a 2019 study in the *Lancet*.

Globally, more than 11 per cent of deaths are caused by sepsis, a life-threatening blood infection, and the chances of this occurring rise when giving birth through a caesarean section or with assistance from forceps or a suction cup. Antibiotics are routinely used as prophylaxis during C-sections, as with other surgeries, but 15 per cent of women still get an infection, WHO figures show. A 2019 trial showed that antibiotics could also be used to further prevent infections during assisted births, by as much as 8 per cent, but this suggestion now comes with the fear of more resistance developing, as it would mean exposing tens of thousands more women to antibiotics. So a seemingly positive outcome and suggestion from a trial on more than 34,000 women brings complex decisions balancing risk and benefit because of the situation we find ourselves in today.

The risk of hospital

The hospital is a setting where resistance has become centre-stage and is currently the main place where infections that are resistant to common antibiotics are not only surfacing but evolving and spreading. The infamous MRSA is a prime example. This is in part due to the scale, the sheer frequency at which infections occur inside a hospital, or in a healthcare setting in general. The WHO states that hundreds of millions of infections occur in such environments each year, with 7 per cent of all hospital patients in developed countries likely to acquire at least one infection (10 per cent in developing countries), the most common being urinary tract infections. Among intensive care patients this number goes up to 30 per cent. A group of bacteria known as the ESKAPE pathogens are the leading cause of hospital infections worldwide, the acronym standing for *Enterococcus faecium, Staphylococcus aureus, Klebsiella pneumoniae, Acinetobacter baumannii, Pseudomonas aeruginosa,* and *Enterobacter* species. These bacteria also show high degrees of drug resistance, for example against the common antibiotic Methicillin, with MRSA (standing for Methicillin-resistant *Staphylococcus aureus*) now present in most countries. The WHO reports the presence of MRSA in five of its six regions, and the strain has proved to be particularly effective at spreading inside hospitals, resulting in wound infections and sepsis. Strains of *Klebsiella pneumoniae,* which causes pneumonia, urinary tract infections and sepsis, with resistance to two classes of antibiotics – cephalosporins and carbapenems – are now found in every WHO region. These examples are just a small window into the plethora of resistant infections taking over hospital settings, leaving doctors with minimal options to treat them and resulting in previously preventable fatalities. For example, people infected with MRSA are estimated to be 64 per cent more likely to die than people infected with a regular, non-resistant strain of *S. aureus*.

Antimicrobial resistance remains a very real challenge, the European Centre for Disease Prevention and Control (ECDC) declared following the release of 2018 surveillance data, which found that more than half of *E. coli* strains and a third of *K. pneumonaie* strains in Europe showed resistance to at least one antimicrobial drug group. Here, the greatest burden is in Italy, Greece, Romania, Bulgaria and Portugal. With very few new antibiotics in the pipeline, healthcare providers and patients themselves are under pressure to keep the current drugs working. This comes with a complex set of guidelines: preventing infections in the first place with good hygiene; infection control and vaccination where possible; and responsible use of antibiotics, using them only when truly necessary, finishing the course and not sharing the drugs with others. All of which will only slow down the problem, not reverse it, but acting now, immediately, is vital.

Sexual revolution

Drug-resistant bacteria have invaded every aspect of our lives in recent years, including our sex lives: causing a rise in sexually transmitted infections (STIs) that are, or were, curable with antibiotics but which health professionals are now struggling to find treatments for. Of the 30 pathogens known to transmit through sexual contact, just four are curable and of these, three are currently surging in developed populations; unsurprisingly these three are bacterial infections: gonorrhoea, chlamydia and syphilis.

Rates of these infections are high in low and middle-income countries, with the WHO predicting global rates each year of 131 million people infected with chlamydia,

78 million with gonorrhoea, and 5.6 million with syphilis. But it's in high-income countries that trends have changed and resistance in particular is becoming an issue, with some cases untreatable by all known antibiotics.

Let's start with the United States, where cases of gonorrhoea, chlamydia and syphilis reached an all-time high in 2018, according to the US Centers for Disease Control and Prevention. Cases of the most infectious stages of syphilis rose almost 20 per cent between 2017 and 2018, from 30,000 to 35,000 cases respectively. Gonorrhoea increased 5 per cent to over 580,000 cases and chlamydia increased 3 per cent to more than 1.7 million cases. These are far from small numbers. Public Health England reported very similar increases over the same time period: a 5 per cent and 6 per cent rise in syphilis and chlamydia, respectively, but an overwhelming rise in gonorrhoea infections of 26 per cent, from around 44,000 to 55,000 cases in England alone. A similar trend is also found in Australia, where chlamydia cases topped 100,000 in 2017 and gonorrhoea went over 28,000.

That's a lot of numbers, but the point they make is that the rise in STIs is global – as are the reasons for their increase: decreased condom use, poor detection rates, poverty and drug use limiting access to sexual health services, and cuts to those services. Infections are typically more common among men and vulnerable groups, such as homosexual men, but all groups are nonetheless seeing a rise, causing a major public health concern. If left untreated the infections affect fertility and can cause ectopic pregnancies, miscarriage, pelvic inflammatory disease and increase the risk of contracting HIV by two to three times. But the problem is now compounded by the emergence of drug resistance, which could leave people untreated if drugs are no longer effective; a scenario that's increasingly impacting treatments for gonorrhoea.

The World Health Organization has reported widespread resistance to common antibiotics among cases of gonorrhoea worldwide, identifying problems in 77 countries, many of which are already using the last-resort antibiotic, extended-spectrum cephalosporins (ESCs). Resistance has even emerged to this drug. 'The bacteria that cause gonorrhoea are particularly smart. Every time we use a new class of antibiotics to treat the infection, the bacteria evolve to resist them,' Dr Teodora Wi, Medical Officer within the Human Reproduction team at the WHO, said in 2017. The agency's Global Gonococcal Antimicrobial Surveillance Programme reports that the first-line treatment for gonorrhoea, ciprofloxacin, is all but useless, with 97 per cent of countries reporting resistance. The next option, azithromycin, is a close second at 81 per cent, hence most countries are already in the last-chance saloon, using ESCs in desperation.

In 2016, an outbreak of gonorrhoea in Hawaii resulted in seven people being diagnosed with the disease and while the first-line treatment worked on them, it worked reluctantly, as the bacteria were showing some signs of resistance, laboratory tests confirmed. In 2018, England was faced with its first case of seemingly untreatable gonorrhoea in a man who picked up the infection while on holiday in South-East Asia. The primary antibiotics to kill the bacteria were ineffective, causing doctors to try ertapenem, an antibiotic often used to treat severe infections of the skin, lungs, stomach, pelvis, and urinary tract, which, thankfully, proved to be effective. Two more cases of gonorrhoea resistant to the main drugs used against it were reported in two females the following year, with one contracting the infection in the UK and the other elsewhere in Europe. Other countries now fear they will see the same outcomes in future cases of not only gonorrhoea but also chlamydia and syphilis, where resistance has been seen and will inevitably grow, particularly as so many cases go undetected. In 2016, the WHO revised its guidelines around the STIs, reinforcing the need to treat these STIs with the right antibiotic, at the right dose, and the right time to reduce their spread and improve sexual and reproductive health. And for the people having sex and being potentially at risk, the solution lies in the age-old message of safe sex and getting tested.

'Everyone can substantially reduce their risk by using condoms consistently and correctly with all new and casual partners,' Dr Nick Phin, Deputy Director of the National Infection Service at Public Health England said in early 2019. 'Anyone who thinks they may have been at risk of getting an STI should seek an STI screen at a sexual health clinic.'

As infections go, STIs are possibly the ones people most avoid seeking help for, due to stigma and individuals often believing they are unaffected, but as with everything, it's better to know one way or the other, particularly when treatments are becoming problematic.

WHEN ANIMALS ATTACK 7

In 2013, a family member of a 50-year-old American woman found a large, red, circular rash on her back. It was very large, around 16 centimetres (6¼ inches) in diameter, with a deeper red colouring in the middle, like a bullseye. The woman, from the state of Virginia, had felt a small scab on her back two days earlier, and had a feeling something unusual was going on as she had felt ill ever since. The site of the scab felt like it was burning, as did her body, reaching temperatures above 39°C/102°F (normal body temperature is around 37°C/98.6°F). The day after discovering the rash, the woman developed joint pain, was constantly thirsty and her fever peaked even higher, topping 40 degrees, forcing her to seek medical attention.

It would take a few weeks to ultimately confirm what the woman was suffering from, but doctors made an educated guess, which would prove to be correct. She had Lyme disease, also known as Lyme borreliosis, contracted from an innocent walk in the woods three weeks earlier. A tick carrying the *Borrelia* bacteria that cause the disease had bitten her without her knowing. Not all ticks carry the infectious bacteria, just ones that have recently bitten animals infected with the disease, such as deer, mice and squirrels. The risk of Lyme disease has increased over time, with the disease now present in the United States, most of Europe and forested areas in Asia. Around 30,000 cases of Lyme disease are reported to the US Centers for Disease Control and Prevention each year, but the agency estimates as many as ten times that number may get the disease annually. In Europe, it's estimated that 360,000 people have contracted the disease over the last two decades.

The case of the woman from Virginia highlights the ease with which Lyme disease transmits, as she had simply gone for a walk in the woods, something she may do regularly, and had not felt a tick embed into her skin as it bit her. People rarely do. But her case also highlights another crucial point about infectious diseases today – the sheer number of them we contract from animals, either directly or through vectors such as ticks or mosquitoes. These pathogens humans share with animals, known as zoonoses, are responsible for more than 60 per cent of the infectious diseases we see in humans and, importantly, some of the worst epidemics and pandemics we have experienced to date: HIV, Ebola, swine flu. It's estimated that three out of every four new or emerging infectious diseases in people are contracted from animals, mostly wildlife, meaning global health experts must keep a close eye on the animals we find ourselves in the company of. 'All diseases that are endemic today in humans, likely came from an animal,' says Dr David Heymann, Professor of Infectious Disease Epidemiology at the London School of Hygiene and Tropical Medicine, adding that if we look at the diseases we know today, they probably got into human populations from an animal, circulated for a while and became endemic. Heymann explains that there are typically three pathways that can occur when an animal infection reaches humans: the first is that an infection enters some humans and goes no further, like rabies; the second is that an infection emerges and spreads between people but doesn't prolong, like Ebola or avian influenza; and the third is when a disease goes into populations and becomes endemic, like HIV.

Today, the boundaries between animals and humans are dissolving, making the crossover of infections more common. Human populations are growing in size and people are moving further into new or unknown territories; communities are becoming more dense and more mobile; animal farming practices have become both wider in scale and more poorly managed and monitored; forests are being cleared, bringing humans closer to the wildlife within them; and many populations are increasingly relying on wild animals, or bush meat, for food. These factors all play a part, for example, in how Ebola outbreaks begin: people enter forests and come into contact with animals infected with the virus, promptly contracting it through contact with the animals' bodily fluids or by consuming their meat.

A more layered example is yellow fever, a viral zoonotic disease that usually affects hunters who go into the deep forest and get bitten by wild mosquitoes carrying the virus, explains Sylvie Briand, Director of the Infectious Hazard Management team at the World Health Organization. Infection with the yellow fever virus causes headache, jaundice (hence yellow), muscle pain, fever, nausea, vomiting and/or fatigue, and in some cases becomes more severe to cause haemorrhage. Yellow fever is endemic in tropical areas of Africa and Central and South America, but the disease is easily preventable as it has a highly effective vaccine that provides lifelong immunity. In Angola, in southern Africa, the growth of the mining industry has changed the epidemiology of the disease. Roadbuilding in the forest, for mining purposes, is making the contact between wild disease and human communities much more intense, says Briand. In 2016, this fuelled a huge outbreak that reached the country's capital, Luanda. 'Probably some workers got the disease while working in the forest, came back to the city and it started an outbreak,' says Briand. The result was more than 1,100 suspected cases, at least 800 of which were in Luanda, and over 160 deaths. Some Chinese workers also unwittingly took the disease to their homeland, a country that had never before faced yellow fever. Eleven cases were reported, but the government responded quickly and the virus did not spread extensively. Mosquito populations also weren't adequate enough in that particular region to continue spread of the disease, explains Briand. But if they had been, 'it could have been a major outbreak,' she says.

Diseases originating in wildlife represent a significant threat to global health, with one 2017 paper adding that zoonoses threaten security and economic growth, making them a public health priority. A prime example of the potential damage they pose is the 2003 SARS pandemic, thought to have been present in civets before the virus adapted to be able to spread between humans, infecting more than 8,000 people worldwide. The ongoing HIV epidemic, resulting in almost 38 million people living with HIV in 2018, is a result of a virus first found in monkeys that crossed over to infect humans. Lassa fever is an example of a seasonal zoonotic disease that affects many West African countries during the winter months (December to March) each year.

Infections spread through contact with food or objects contaminated by the urine or faeces of infected rats, and poorer members of society are at most risk as they are more likely to come into contact with rats scavenging for food in their homes. Molecular testing suggests the haemorrhagic disease has been circulating in Nigeria for more than 1,000 years.

The examples go on; MERS-CoV, Nipah virus, monkeypox and rabies are also zoonoses. Even our household cats carry the risk of toxoplasmosis, a parasitic disease affecting pregnant women and people with suppressed immune systems: this can occur if cats eat infected rodents or birds and then shed cysts in their faeces, which humans then come into contact with either directly or through soil containing the faeces.

The biggest challenge is that even today, little is understood about how zoonoses spread and develop, particularly as most of the experts in the field are kept busy reacting to outbreaks that have already emerged, leaving little time to understand or predict new ones. Depending on the infection, pathogens typically cross over between animals and humans either directly, through contact with blood and bodily fluids; indirectly, coming into contact with germs left behind by animals, for example in soil or chicken coops; through vectors, like mosquitoes and ticks; and by eating contaminated food, such as unpasteurized milk or undercooked meat from infected animals. But this is a broad summary and much finer details are needed about transmission routes and cycles, and the pathology and microbiology of emerging zoonoses, for us to truly understand the early signs of an impending outbreak or to predict one before the signs even begin. For this reason, the majority of the 'priority diseases' named on the WHO's research and development blueprint are zoonotic diseases, such as Nipah and Crimean–Congo haemorrhagic fever, included on the list due to their epidemic potential and the need for greater research into them.

But another way to stay ahead is by greater collaboration, says Heymann, through a 'one health' approach. 'The one health movement has developed in order to make sure that animal and human health people are working together,' he says. Effort by just one sector cannot prevent or eliminate the problem, the WHO states, highlighting, for example, that rabies is targeted by vaccinating dogs (which carry the virus) and that 'information on influenza viruses circulating in animals is crucial to the selection of viruses for human vaccines for potential influenza pandemics'. An example of this in practice is the UK's Human Animal Infections and Risk Surveillance (HAIRS) group, which includes representatives from eleven authorities, including governments, health agencies, the Department for Environment, Food and Rural Affairs and the Food Standards Agency. The group have met monthly since 2004 to identify emerging and potentially zoonotic infections which may pose a threat to the UK. 'They look at all emerging infections occurring in the world and evaluate the importance to the UK, the risk to the UK,' adds Heymann. 'If it's something they're concerned about they develop guidelines.' Monthly documents are then published, summarizing 'notable incidents of public health significance', such as the currently ongoing Ebola epidemic in the Democratic Republic of Congo or dengue and Zika cases being reported in Europe. This collaborative approach on a global scale is what it will take just to keep up with the continuous onslaught of infections, let alone be ahead of the game, but experts believe we can get there, and in the grand scheme of things it's the diseases we already know about, like Ebola or influenza, that are the most common concern, helping teams to develop informed strategies to control them.

Ebola

Ebola was once a disease known mostly to rural communities in a select few countries in East and Central Africa: Sudan, Uganda, Gabon and, most predominantly, the Democratic Republic of Congo, where the virus was first discovered and the majority of outbreaks have since occurred. The haemorrhagic disease is first spread to people from infected animals, such as bats or monkeys, but can then spread from human to human through contact with bodily fluids, such as blood or saliva, causing a severe, often fatal, disease. Symptoms include vomiting, diarrhoea, impaired liver and kidney function, and internal and external bleeding. Fatality rates vary from 25 per cent to as much as 90 per cent.

The 1995 film *Outbreak* increased awareness of haemorrhagic diseases somewhat, depicting a scenario where an Ebola-like virus reached the United States, creating a fear factor around the lethal disease and its potential to cause extensive damage beyond Africa. But until recently, Ebola was still largely considered a disease of rural Africa, local to the continent's forested regions. That is, until December 2013, when an eighteen-month old baby would change the worldview, as an index case bringing Ebola to a new region, West Africa, and an unsuspecting population, where the virus would go on to infect tens of thousands of people, soon eliciting global panic.

The baby boy, whose name has not been revealed, developed a mysterious illness that caused him to develop a fever, black stools and vomiting on 26 December 2013 in the village of Meliandou, Guinea, located near the country's borders with Liberia and Sierra Leone. He died two days later and three weeks after that, several of his family members developed similar symptoms and also promptly died. Midwives, traditional healers and staff at the nearest hospital who treated the baby also died. A chain reaction had kicked off, soon infecting members of the boy's extended family who had either taken care of the infirm or attended their funerals, reaching four sub-districts within weeks and the capital city, Conakry, on 1 February 2014. At first it was considered a mysterious disease that had spread silently, knocking down everyone in its path for months as district health officials scrambled to understand what it was. The country's Ministry of Health raised an alert on 13 March 2014 and the mystery was finally solved, naming the outbreak as Ebola virus disease nine days after that. This was the first time Ebola had reached West Africa, meaning it took officials a while to realize what they were dealing with, allowing the contagious disease to spread across Guinea, then the region, infecting over 28,000 people. Liberia and Sierra Leone would prove to be the worst hit, with more than 10,000 and 14,000 cases respectively.

The source of the baby boy's infection isn't clear, but the likely cause was contact with an infected animal, according to the World Health Organization. Forest clearing around the small village of Meliandou, for mining and timber, brought human and animal populations closer together, and experts suggest the baby became infected while playing in his back yard. Such exposure to forest wildlife is the first of many factors that fuelled the explosion of Ebola cases that would follow. Weak surveillance systems and public health infrastructure meant the outbreak took months to identify, delaying any control efforts; the regular movement of people across the borders of Guinea, Sierra Leone and Liberia meant the virus spread easily to new communities and, for the first time, reached dense, urban communities. Inexperience with the disease resulted in poor understanding and misperceptions among affected communities, causing challenges in controlling transmission and finding cases, while inadequate infection control in healthcare facilities and hospitals led to a large number of infections among health workers. 'It was the first time we had an outbreak at the border ... So immediately we had three countries infected,' explains Sylvie Briand.

By July 2014, six months after the first case, the outbreak had spread to the capital cities of all three countries, providing vast opportunities for transmission. The following month, more than 1,700 cases had been reported and a public health emergency of international concern (PHEIC) was declared by the International Health Regulations Emergency Committee. 'A coordinated international response is deemed essential to stop and reverse the international spread of Ebola,' the committee said. But Briand highlights that this declaration, while crucial to get extra resources to contain the outbreak, brought problems of its own. 'As soon as we declared a PHEIC, there was border closure and the countries went from health crisis to a humanitarian crisis,' she says, explaining that there were then no flights and no food. 'We learned from that.'

The international response was unprecedented. Controlling the outbreak would take millions of dollars, unimaginable levels of resources and teams on the ground, and the trialling of an experimental vaccine that had long been ignored because it wasn't considered to be essential.

The vaccine, called rVSV-ZEBOV, created by pharmaceutical company Merck, Sharp & Dohme, was used on compassionate grounds, using a ring strategy where people who had come into contact with infected persons would receive the vaccine and be monitored to see if they developed the disease. The effectiveness of the vaccine was to be analysed in the field, the only real option for infections like Ebola that are not endemic and therefore do not otherwise have people to trial a vaccine on. The use of an experimental vaccine in this way, however, was in itself a wake-up call to the public health community to prioritize vaccines against seemingly rare diseases in today's globalized world, particularly as no cases of Ebola developed among people who were vaccinated. The trial found rVSV-ZEBOV to be 100 per cent effective. 'While these compelling results come too late for those who lost their lives during West Africa's Ebola epidemic, they show that when the next Ebola outbreak hits, we will not be defenceless,' Dr Marie-Paule Kieny, the WHO's Assistant Director General for Health Systems and Innovation, said at the time. But while the vaccine was highly effective, it was improved infrastructure and services, carefully planned policies and extensive engagement with affected communities and the leaders of those communities that played the biggest part in curtailing this outbreak. With a contagious disease like Ebola, when people are afraid of being quarantined or believe any myths around the transmission of the disease, infections will continue.

It took time to build trust within the communities at risk, but two years later, in March 2016, the WHO announced the outbreak was no longer a PHEIC and three months after that, in June, the largest-ever outbreak of Ebola would finally be declared over. More than 11,000 people had died and the outbreak had reached ten countries, the epicentre by far being the West African countries of Guinea, Liberia and Sierra Leone. Thirty-six cases and fifteen deaths occurred in the other seven countries, the majority in Nigeria, with four cases in the United States.

Since the first known outbreak of Ebola in the DRC in 1976, more than 20 outbreaks had occurred prior to 2014, some large – affecting hundreds of people – but most of them small, affecting tens of people in a particular village or community. The lethality of the virus meant it soon died out. The circumstances in West Africa, however, showed the world that Ebola could in fact explode and one day arrive on their doorstep. The World Health Organization was widely criticized for waiting too long to appreciate the urgency of the situation in 2014.

Since 2014, there have been three outbreaks of Ebola, all in the DRC, a country experienced in controlling the disease. Familiarity meant alarms were not raised for the first two outbreaks, which affected eight and 54 people respectively. They were kept under control. In fact, another outbreak had also occurred in the DRC during the one in West Africa, in 2014, and was swiftly curtailed. But in August 2018 we learned that even in a country well versed in Ebola, if the virus strikes in a new part of the country, where local communities and leaders don't know what they're dealing with, another explosion will occur. And it did, in the eastern provinces of North Kivu and Ituri.

Though beginning in summer 2018, the North Kivu/Ituri outbreak surged almost a year later, in April 2019, reaching a peak between June and August, when between 75 and 100 cases were reported each week, and the virus crossed over the border into neighbouring Uganda, causing four cases there, as well as reaching the city of Goma in North Kivu, which has a population of 1 million, according to Médecins Sans Frontières (MSF). The arrival of the virus in a big city led to the outbreak being declared an international concern, a PHEIC. Fortunately, just two cases occurred in Goma, but the spread of Ebola to a third province, South Kivu, meant it was not under control. Local, national and international teams have been on the ground since the beginning of the outbreak, improving treatment facilities, tracing contacts and administering vaccines, which included the use of a second experimental vaccine by Johnson & Johnson in November 2019, in addition to the rVSV-ZEBOV vaccine by Merck, which has been used since the outbreak began.

But a number of social and political factors continue to block control efforts. Populations in the region are very mobile, fearful communities do not always disclose cases or contacts, and civil unrest in the region has seen health workers and vaccinators attacked. By December 2019, there were 386 such attacks, resulting in seven deaths and 77 injuries among healthcare workers. 'Ebola was retreating and now it is likely to resurge,' the WHO's Mike Ryan, Head of Emergencies, said at the time. MSF described the outbreak as unpredictable, which, when added to the extreme difficulty in accessing some of the affected regions – due to both infrastructure and security – meant there was a 'lack of visibility on the epidemiological situation'. People still weren't clear on the extent of the outbreak.

By the beginning of March 2020, eighteen months after the outbreak began, more than 3,440 cases had been reported, with over 2,260 deaths – a fatality rate of 66 per cent – and the outbreak was still ongoing, albeit with much less vigour. In fact by March 2020, new cases had not been reported for two weeks, which brought gentle hope to those leading control efforts. This is the DRC's tenth Ebola outbreak, with the previous nine affecting less than 10 per cent of the number affected in the current outbreak. This tenth outbreak for the DRC has become the second largest outbreak of Ebola on record and, extraordinarily, the country has simultaneously faced a record measles outbreak, the largest in the world in fact, far outweighing the Ebola numbers. Measles is affecting all of its provinces, with over 250,000 suspected cases and more than 5,000 deaths as of December 2019. 'While the Ebola outbreak in the DRC has won the world's attention and progress is being made in saving lives, we must not forget the other urgent health needs the country faces,' Dr Matshidiso Moeti, WHO Regional Director for Africa, said at the time. Low immunization rates and high levels of malnutrition are believed to be underlying factors for the surge in measles cases and related deaths. But the scenario as a whole proves a vital point about infectious diseases: that even when experts think they know what to expect, or have something monitored or under relative control, an infection – be it Ebola or measles or anything else – always has a hand it hasn't played yet.

The many faces of flu

Influenza is underestimated. As a seasonal infection that strikes most of the world's population, it's an expected part of winter life. People with a slight temperature and runny nose will often say they have the flu, though a true infection would leave them bedridden for at least a week. Many actually are bedridden for a week or more but know they will recover; many receive a newly designed vaccine each year, targeting that year's strains, in hope of damping the effects of the virus to keep them healthy during the colder months; some keep a watchful eye, almost expecting it to strike at any minute. Parlance around influenza is commonplace in our society, but it could be said that the majority of people don't appreciate the potential lethality of an infection. If it strikes the elderly or young, or if a pandemic strain emerges, people know to worry and seek immediate help. But for anyone aged in between who contracts the seasonal virus, it's assumed bed rest and liquids will see them through. Influenza, however, is a complex beast.

Seasonal influenza causes extensive damage to a population's health. The annual epidemics are estimated to cause about 3–5 million severe cases of disease each year, and between 290,000 and 650,000 deaths. The United States sees between 12,000 and 61,000 deaths annually, and Europe 15,000 to 70,000. The wide range in these figures is because there are different types and subtypes of the virus and each year new strains of these circulate, and their virulence, their potency, is hugely variable. For example, the 2017–18 season was the most damaging in recent years, with two flu subtypes circulating and the vaccine being less effective than usual, causing more than 44 million cases and 61,000 deaths in the United States and an estimated 152,000 deaths in Europe. Severe cases and deaths are predominantly among high-risk groups, such as the elderly, immunocompromised (people with weakened immune systems) or chronically ill, but it's fair to say the harm caused by influenza is far greater than people realize. The number of deaths worldwide is often greater than deaths from malaria, for example.

Before going further, and before exploring the P-word – pandemic – it's important to understand the basic biology of influenza. The airborne respiratory virus has four types: A, B, C, and D, with the first two being the cause of seasonal epidemics and type A causing the greatest burden. Type A further divides up into subtypes, based on the combinations of two proteins found on its surface, haemagglutinin and neuraminidase, examples being A(H1N1), which circulates every year, but a version of which was behind the 2009 swine flu pandemic. Type C is less common and when it does arise, causes mild illness, meaning it is not a public health priority, while type D is known to infect cattle, not humans, for now. A mix of type A and type B viruses circulate at any one time and they change over time, due to genetic mutations and gene reassortment, meaning we see different strains of the virus types each flu season. This explains why people need to get vaccinated every year, as a new vaccine will have been developed to match the strains currently circulating. In fact, the vaccine is updated twice a year, for flu seasons in the northern and southern hemisphere, which occur six months apart. The WHO Global Influenza Surveillance and Response System, a system of National Influenza Centres and WHO Collaborating Centres around the world, continuously monitor what's circulating and determine the vaccine's composition.

The majority of infections are uncomplicated, meaning people recover after a few days or weeks. Symptoms in these cases include fever, tiredness, a runny nose, headache and muscle pain. Some studies estimate that as many as 75 per cent of infections are asymptomatic. But away from these milder infections, a great many severe infections and fatalities occur, as earlier numbers show. More severe infections can cause pneumonia and inflammation of heart muscles and the brain, exacerbated by any underlying conditions a person may have, such as diabetes or heart disease. As a result, they typically affect the elderly, pregnant women and the chronically ill, which is why these groups, as well as children, are prioritized to receive vaccines each season.

The flu vaccine itself is complex and its effectiveness extremely variable, because it depends if the correct strains for the upcoming season have been chosen, based on what is circulating in advance and when the vaccines need to be designed and made. Any protection individuals may have against some types or subtypes, from previous infection or vaccination, do not confer immunity to other types and subtypes. On average the flu vaccine reduces the risk of developing flu symptoms by between 40 per cent and 60 per cent in the population as a whole. During the more severe epidemic of winter 2017–18, the vaccine was just 40 per cent effective, according to the US Centers for Disease Control and Prevention. This is overall protection against all flu strains, with percentages varying for each strain included in the vaccine. For example, the 2017–18 vaccine protected against A(H3N2), A(H1N1) and influenza B viruses and reduced risk of disease from each type by 25 per cent, 65 per cent and 49 per cent respectively. Vaccination remains the best means of prevention though, in addition to good hygiene and isolation once symptoms develop. Treatment is also an option, particularly for people at risk of severe disease, using antivirals such as oseltamivir, sold as Tamiflu. These can also be used as a means of prevention in some circumstances.

The endless cycle of influenza leaves no time to be complacent. Prevention efforts are continuous and responsive and require extensive resources, leaving significant economic burden. In the United States, for example, one study estimated the annual economic burden to be over $11 billion, accounting for costs to both healthcare and society from not only the severe cases and deaths, but the mild to moderate cases which result in time off work and losses to productivity. This significant cost is for a regularly occurring infection that we see coming each year, so imagine the cost when it comes as a surprise. When that happens, the global cost is estimated to be almost six times this figure, at $60 billion a year. It is in this scenario that populations truly fear the damage influenza can bring. And so we reach the P-word, the event in which the spread of a novel influenza virus across the globe is most feared.

In 1918, more than 50 million people died from an H1N1 influenza pandemic that spread across the world. In 2009, the same flu subtype returned. In both cases it emerged from pigs, but in 2009 times were very different; healthcare was a lot stronger, people were healthier overall and the virus strain was less virulent than experts had predicted. Nonetheless, very rough estimates suggest more than 280,000 people died worldwide, close to the minimum number that die from seasonal strains each year. But novel strains that underlie a potential pandemic spark worry because they are a surprise, they are unknown, and teams have not predicted their arrival so as to at least have a partially effective vaccine available ahead of time. Many countries faced the first wave of the 2009 H1N1 pandemic without a vaccine, because it arrived late, explains Sylvie Briand of the World Health Organization. 'Then when it arrived, some countries had too much, others had none,' she says, which led to those with more donating their extra supplies to the WHO, by which point it was 'too little too late'.

Influenza viruses that go on to cause pandemics are always type A and so invariably have a subtype designation referencing the proteins found on the surface (such as H1N1). They are able to cause pandemics because of large genetic changes the virus undergoes, due in part to the fact this virus type is also found in birds and swine. Influenza A is transmitted interchangeably between humans and swine, and also from birds to humans. Surface proteins from avian or swine influenza viruses can become incorporated into human viruses, through gene reassortment. This is a common occurrence, but not all reassortments lead to viable viruses and even ones that are viable may not be able to infect humans, or enable ongoing human-to-human transmission. They may also be a strain to which humans harbour some immunity. A perfect storm of factors must come together, but on occasion they will and this is when a pandemic strikes.

In 2009, H1N1 crossed over from swine to readily infect humans. Prior to that, in 1997, H5N1 crossed over for the first time from birds, causing the first human case of avian influenza (bird flu). H5N1 went on to infect people around the world, with over 250 cases and 150 deaths across at least 55 countries, but it was not virulent enough to cause a pandemic. In 2003 it re-emerged, reaching many countries in Asia, but with fewer cases: 23 confirmed cases, of which eighteen died, and around 100 suspected cases, according to the US Centers for Disease Control and Prevention. Avian influenza remains a regular threat among birds, with farmers worldwide monitoring and culling their flocks accordingly, and human cases still occur most years, albeit in cases totalling single figures, among people in prolonged close contact with the birds, most often in China. But its emergence and re-emergence left countries scared, leading to intensive work on preparedness, explains Briand. 'Countries were really frightened to have a true pandemic,' she says.

Be it a crossover from pigs or from birds, global health experts around the world know the chances of another pandemic strain emerging are high. There is no reason why one of the many gene reassortments that occur will not result in a virus with the potential to infect thousands worldwide, so health authorities must increase their capacity to predict and their ability to react. But whether it's a pandemic or seasonal influenza, it's all a guessing game using the information to hand, with everyone hoping they guess right.

'I'M NOT GOING ANYWHERE' 8

In the early part of the twentieth century, one particular disease was considered to be a major public health problem and it's probably a disease you've never heard of: yaws. A bacterial disease affecting the skin, bone and cartilage, yaws spread through contact with minor wounds and primarily affected children, leaving them chronically disfigured. The disease is part of a group of chronic bacterial infections commonly known as the endemic treponematoses, which includes syphilis.

Research shows that in the first half of the twentieth century, yaws was a predominant health concern in the majority of countries worldwide. In 1936, for example, it constituted more than 62 per cent of infectious diseases being treated at health facilities in Ghana, and more than 47 per cent of infectious diseases being treated in Nigeria the year before. To really highlight the point, it was estimated that 160 million people were infected worldwide in 1950. Therefore, in 1949, one year after the World Health Organization was formed, the World Health Assembly resolved to control yaws, using the newly developed benzathine penicillin, which could cure people of the disease with a single injection. Three years later, in partnership with UNICEF, a campaign began to control and eventually eradicate the disease, using mass treatment campaigns in 46 countries.

If you had never heard of the disease before now, you'd be forgiven for thinking the campaign was successful and that yaws was indeed eradicated. But alas, it was not. Numbers did decline dramatically, by 95 per cent within twelve years, from 50 million in 1952 to 2.5 million in 1964, and the campaigns helped establish primary healthcare in the countries targeted. But then there was a U-turn. A change in strategy was introduced to tackle the last 5 per cent of cases: incorporating the surveillance and control of yaws into the newly established primary healthcare system. This simply didn't work. The health infrastructure wasn't strong enough yet and so both surveillance and commitment waned. The disease began to re-emerge by the 1970s. Cases were estimated to be 460,000 in 1995 and infections still persist today. Yaws is known to currently be endemic in fifteen countries with three others reporting suspected cases and many more needing to be assessed.

But yaws is no longer considered a public health priority by many policymakers, with the result that commitment is still lacking. There has been some progress since the 1960s, however, firstly with India declared to be free of yaws in 2006, having reported its last case in 2003, after seven years of national efforts to end the disease. It was also discovered, in 2012, that a single dose of the oral drug azithromycin cured people with the disease, offering an alternative to the penicillin injection, making treatment simpler and more accessible. Use of the new drug revived interest in eradicating the disease worldwide and a new programme was launched, hoping to eradicate yaws by 2020. This goal has not been met and has been pushed back to 2030, along with the target for Guinea worm, as part of the roadmap for all neglected tropical diseases. But the hope of eradication, or at least elimination, remains on the cards; and herein lies a crucial point about the future of disease control.

When it comes to disease control the ultimate aim, or ambition, is to wipe out the disease at hand. Rid the world of it completely. But in reality, this will almost always be a pipe dream. The world came together and did it for smallpox, but has struggled ever since, despite launching programmes for yaws (before smallpox), malaria, polio and Guinea worm. The last two may be very close – but they have been so for years and are likely to be for many more. This then begs the question: is it enough to just get close to eradication? To bring case numbers for a disease down by so much that millions of lives have been saved, and will continue to be saved, but the disease will still lurk around in a minority? Because answering yes to these questions is the realistic option, most experts agree, and one that may actually bring some success to global health programmes.

'Elimination doesn't mean getting it to zero, it's by definition a much easier target than zero,' says Dr Donald Hopkins, who led efforts to eradicate smallpox. He explains that the general goal should be to reduce suffering. 'The important thing is what's the best way to reduce the greatest amount of suffering. In some instances it will be just control, in other instances it will be elimination; in a few instances it will be eradication,' he says.

A long list of diseases are currently targeted for elimination, mostly neglected tropical diseases, such as rabies or Chagas disease (more on this disease below), that affect the poorest populations globally. Depending on the disease, elimination is being focused on either globally, regionally or nationally in certain countries; and the definition used for elimination is also bespoke: it could mean limiting transmission within a particular area, ending deaths from a disease, or reducing numbers to a certain number globally. For example, the 'Zero by 30' plan aims to end human deaths from rabies by 2030, specifically rabies caused by dogs, which accounts for 99 per cent of cases. This is being rolled out by a collaboration of four global agencies: the World Health Organization (WHO), Food and Agriculture Organization of the United Nations (FAO), World Organisation for Animal Health (OIE) and the Global Alliance for Rabies Control (GARC).

Chagas disease, also known as American trypanosomiasis, is a parasitic disease spread through the urine or faeces of the triatomine bug, in which the bugs bite people and urinate or defecate near the bite, so the parasites enter the bloodstream if the person then wipes the skin area near the bite. Chronic infections lead to heart problems, but there is a drug and infections are curable if treated early. The disease is predominantly found in the Americas and because the parasite behind the disease, *Trypanosoma cruzi*, is also common in wildlife, the notion of eradication is off the cards. Instead, goals for disease control are to eliminate human transmission and to ensure early access to healthcare, as it's the most remote, poor communities that typically get infected. The list goes on to include at least seven more diseases with elimination programmes underway, one of them being leprosy, touched on in chapter 3, for which one elimination target was met in 2000 and another was set in 2016.

Leprosy was targeted for elimination as a public health problem, which meant bringing prevalence of the disease down to less than 1 case per 10,000 people, and this was met. As we saw earlier, case numbers today are approximately 0.2 per 10,000 people, with almost 80 per cent of them in three countries: India, Brazil and Indonesia. In 2016, WHO launched its Global Leprosy Strategy, which focuses on factors including reducing infections among children, with the goal of zero disabilities among children, and stopping all legislation that allows discrimination of people with leprosy.

The point here is that these programmes have made significant strides towards bringing the burden of disease down to minimal levels – almost zero in some cases or regions – reducing mortality and illness among affected populations, which is an achievement for global health. Such programmes have largely targeted diseases of the poor – neglected tropical diseases, as they're called. The challenge now is to ensure control efforts and reduced case numbers are maintained because the pathogens behind the diseases are still around, whether in animals or in a miniscule number of people. 'The problem is that many people don't understand the difference between eradication and elimination,' says Dr David Heymann, Professor of Infectious Disease Epidemiology at the London school of Hygiene and Tropical Medicine. 'They think that they're going to get rid of the programme and not have to do any more care of people with leprosy or lymphatic filariasis or whatever, so there is a misunderstanding.' The need for commitment is permanent.

But there are two diseases, one of which has been around for centuries and the other just a few decades – at least known among humans – that have had a complex history in terms of control: malaria and HIV/AIDS. There has been significant progress in reducing the burden of these infections, but they continue to affect millions worldwide today, despite large programmes and resources to control them. Malaria affects more than 200 million people each year, and HIV more than 37 million, with HIV moreover leaving these millions with a lifelong infection. Bespoke programmes and initiatives are underway to 'end them' but the complex biology and pathology of the parasites and viruses means the journey has been extensive and there remains a long road ahead – even to reach elimination.

No more malaria

There has long been hope of eradicating malaria – for almost a century in fact – and that hope remains today, with a global technical strategy for malaria in place that hopes to deliver a malaria-free world. The strategy lays out a plan to reduce infections and deaths by 90 per cent and eliminate malaria in at least 35 countries by 2030, and experts recently argued that eradication could be achieved by 2050, including long-term funder Bill Gates, who has said malaria eradication is possible in his lifetime. In fairness, the disease has been eliminated in many regions of the world, but where it remains, the burden is prolific. This is largely in low-income countries, among the poorest subsets of those populations. It is therefore also fair to say that eradication is an ambitious target. 'Eradicating malaria has been one of the ultimate public health goals,' writes the World Health Organization's Director General, Tedros Adhanom Ghebreyesus, in a 2019 paper in the *Lancet*. But he promptly adds, 'It is also proving to be one of the greatest challenges.' Other experts add to this, stating that 'the challenges remain formidable'. So why do people think it is possible?

Malaria is a potentially fatal disease caused by infection with plasmodium parasites that spread through the bite of female *Anopheles* mosquitoes, as the females feed on blood to nurture their eggs. The parasites have complex life cycles inside both the mosquito and humans where they develop into different forms, becoming able to infect each host interchangeably. Infection results in fever and headache at first, with symptoms worsening to include organ failure or severe anaemia if people aren't treated within 24 hours, often then leading to death. Children under five years of age, pregnant women and patients with HIV/AIDS are at most risk of developing severe disease if infected. As with most infectious diseases, malaria is a complex disease, aided by great variability in how it plays out in different regions of the world, harnessing different versions of the parasites and mosquitoes.

The desire to control malaria on a large scale, and garner enough support to ensure resources to do it, began during the Second World War when infections were striking down large numbers of soldiers. Prior to this, efforts had been made to control levels of the disease in the United States, for example, by controlling water levels to reduce breeding sites for mosquitoes and the gradual use of insecticides, and this showed that case numbers could be brought down. Wartime saw the first large-scale use of the insecticide dichlorodiphenyltrichloroethane (DDT) to keep mosquitoes at bay at Allied camps across the world, and control efforts were then also used in training camps across the southern states of the US, where mosquitoes are found, laying the foundation for the creation of the US Centers for Disease Control and Prevention. According to the Malaria Consortium, use of DDT in this way was a precursor to the national malaria eradication programme in the US, which began in 1947. By the 1950s, malaria had been eliminated from most temperate regions of the world, giving hope of global control.

In 1955, the Global Malaria Eradication Programme began, bringing much success, but also much failure. Eradication efforts involved spraying the inside walls of houses with insecticides (as mosquitoes rest here after feeds), treatment with a cheap and effective antimalarial drug, chloroquine, and surveillance for cases. Many countries soon became 'malaria-free' and others saw case numbers drop dramatically. The campaign eliminated malaria from Europe, North America, the Caribbean and parts of Asia and South and Central America, though this decline was again mainly in countries with temperate climates or seasonal malaria transmission. Some countries, such as India and Sri Lanka, saw sharp declines, but then substantial surges once control efforts stopped. Other countries, such as Haiti and Indonesia, saw no significant decline and, astonishingly, most countries in sub-Saharan Africa were left out of the campaign altogether – countries which now have the greatest burden of disease. Ghebreyesus writes that the campaign was 'flawed from the start by leaving out tropical Africa'. The region was largely ignored due to logistical difficulties and the challenges of executing control efforts there. But today, this is where 80 per cent of the 228 million cases estimated worldwide are found.

The eradication programme faced multiple challenges in terms of the tools being used to control the disease on the ground. Drug and insecticide resistance emerged – a significant barrier – and further challenges included wars and massive population movements, difficulties in obtaining sustained funding from donor countries, and lack of community participation, making the long-term maintenance of the effort 'untenable', according to the US Centers for Disease Control and Prevention. Many experts highlight that too much emphasis was put on one single strategy, insecticide spraying, which was impacted by the fast development of resistance among mosquitoes, bringing the realization that 'no single strategy can be applicable everywhere'.

The programme just didn't work and any further hope of eradication was abandoned fourteen years after it began, in 1969. Economic crises in the following decades reduced global support for malaria control even further and by the 1990s, case numbers had gone up again and the burden of the disease had worsened significantly, prompting renewed calls for control efforts worldwide. But control of the disease and a substantial decline in cases didn't really occur until after the year 2000, when a summit in Abuja, Ethiopia motivated African leaders to sign a declaration, the Abuja declaration, committing to try to halve malaria mortality for Africa's people by 2010, in partnership with the Roll Back Malaria initiative. The Millennium Development Goals were also set and the Global Fund to Fight AIDS, Tuberculosis and Malaria was launched.

In the time between giving up on malaria in 1969 and reigniting hope in 2000, valuable research had been conducted on drug and vaccine development and vector control, leading to the development of some effective tools, namely drugs and better mosquito control methods. Combination therapy involving the drug artemisinin is effective at treating the disease today, after former drugs such as chloroquine became ineffective in many regions due to the emergence of drug resistance. The creation of rapid diagnostic tests also meant people could be diagnosed and treated quickly, preventing the onset of severe disease. Control efforts also prioritized malaria prevention, mainly by reducing mosquito bites through the use of insecticide-treated bed nets, but also continued efforts to spray the inside of homes and buildings with insecticides. Preventative drugs that suppress the early stages of malaria infection – targeting the parasites as they first enter the blood cells, blocking them from establishing themselves – were also used. This is known as chemoprophylaxis and continues to be used today in travellers, pregnant women in endemic regions and populations living in areas where malaria is seasonal, blocking surges that typically occur just after a country's rainy season.

Between 2000 and 2015, the malaria epidemic was struck hard and numbers came flying down. One study estimated that new infections fell by 37 per cent and malaria deaths declined by 60 per cent, from an estimated 839,000 deaths in 2000 to 438,000 in 2015. It's believed this progress was largely a result of greater access to mosquito control interventions, particularly bed nets, which were provided to many remote communities. But, perhaps unsurprisingly, the regions with the greatest burden saw the slowest decline over this period: 88 per cent of cases and 90 per cent of deaths still occurred in the WHO African region, for example. Since 2015, numbers have continued to decline, but at a much slower rate, having almost plateaued, data shows.

'We need to step up action, particularly in the hardest hit countries,' said the Director of the WHO Malaria Programme, Dr Pedro L. Alonso during a press conference for the 2019 Global Malaria Report, which estimated that 228 million cases of malaria and 405,000 deaths occurred worldwide in 2018, an almost negligible decline from the

251 million cases and 416,000 deaths estimated in 2017. The first part of this century showed that progress is possible, but 'progress has slowed down, we have stabilised at an unacceptably high level', he said. Before this plateau, the idea of eradication was put back on the table in 2016 and a strategic advisory group on malaria eradication was formed to look into the factors and determinants that underpin malaria. The group accounted for trends that may affect the disease epidemiology, such as changes in agriculture and land use, urbanization, migration and climate, to determine if eradication was worth pursuing. In 2019, they decided it was. 'There are no biological barriers to the concept of eradicating malaria; it can be done,' said Alonso during a press conference with the advisory group in August 2019. However, he then added that the group is currently unable to formulate a precise and reliable plan for eradication, so really the answer is 'watch this space'.

It should be pointed out that it has not been all bad news in recent years: many countries have been certified, or are on track to being certified, malaria-free, with Argentina and Algeria joining the certified list in 2019 and El Salvador and China hoping to get there in 2020. Iran and Malaysia are getting close too, says Alonso. This highlights a particular point about the malaria epidemic, he adds, which is the fact that there are two types of countries: ones with a heavy burden and ones within reach of elimination. Global eradication means tackling the former group, those with the heavy burden, and for them, 'the currently available tools and approaches will not be sufficient to achieve malaria eradication', writes Tedros Adhanom Ghebreyesus. The strategic advisory group analysed the outcome if current tools were scaled up to an 'absolute maximum', which meant above

90 per cent, and found that, despite a dramatic reduction, this would still leave more than 11 million cases by 2050. 'We would still be quite short of having eradicated this infection,' said Alonso. Other experts add that even new tools are not enough, as we need a strategy to integrate them effectively into different health and social systems, which vary greatly across the world, as well as greater surveillance methods and a large injection of money – at least US$34 billion, which would only cover efforts until 2030.

One new weapon recently joined the battle against malaria: a vaccine. Called RTS,S, the vaccine has been shown to reduce malaria infections among young children by 39 per cent in clinical trials and further showed protection during a pilot introduction in three countries in sub-Saharan Africa in 2019. Now, it has completed its full development and over 300,000 children will be immunized every year in the coming five years, explained Alonso. But it is not a vaccine in the usual sense, as it only really provides protection among children for a few years. But it helps. 'Imperfect tools allow us to achieve great impact,' said Alonso. And that sums it up, really. The fight is about throwing whatever you have at the parasite to prevent as many cases as you can and treat any cases that do break through as quickly as possible. All this, tailored for different societies and health systems, will make a dent. Only then will we maybe, just maybe, have some hope.

So, to ask the question again, will malaria be eradicated? Experts believe it can be but whether it actually will be, or how it could be, they can't yet say. As with most diseases, eradication will prove to be a battle we may not win. 'A goal is set and it's worthy to follow that goal,' says Dr David Heymann. But he adds a note of reality that everyone is aware of. 'Elimination is possibly a more worthy target because there are mosquitoes involved,' he says. 'It's very difficult to deal with malaria.'

Getting a handle on HIV

On 24 April 1980, the United States Centers for Disease Control and Prevention (CDC) received word of a patient diagnosed with Kaposi's sarcoma, a rare type of cancer that seizes upon a weakened human immune system. The patient, Ken Horne, was extremely weak and his immune system almost completely depleted. The cause of his cancer was unknown and experts scrambled to try to understand what was going on. Horne would later become known as the first case of Acquired Immunodeficiency Syndrome (AIDS) in the US, and the first case we're aware of anywhere. A year later, five otherwise healthy gay men were diagnosed with a rare form of pneumonia called pneumocystis pneumonia (PCP). They were separate cases across three hospitals and didn't know each other. The CDC knew this was unusual, as this form of pneumonia again only affects severely immunosuppressed patients, and the agency's report from the time suspected they had all been exposed to something through some aspect of their lifestyle. Soon they were getting calls from various doctors reporting similar cases across the country.

An epidemic of AIDS was underway, starting a race to first identify this new disease and then to find out how it was being spread and what was causing it. Only in 1983 did scientists discover a virus was behind it, which they called the human immunodeficiency virus (HIV). These early cases opened the floodgates to reveal hundreds of others experiencing symptoms of AIDS – later described as infection with HIV – not just in the US, but worldwide. They had unearthed a global pandemic that had secretly survived and spread for years, hiding in communities that no one had paid attention to. Now the world had to find them while also finding a way to control the virus as soon as possible.

'One of the main features of this infection is its latency,' says Dr Linda-Gail Bekker, Deputy Director of the Desmond Tutu HIV Center in South Africa and former President of the International AIDS Society. 'Most people don't know they're infected,' she says. People have up to ten years in which they can unknowingly transmit the virus before developing symptoms. 'That's how it just explosively spread.'

Another crucial factor within this explosion, with huge socioeconomic impact, was the fact that HIV took down the 'middle men' says Bekker, the young, active, working-age population, people of reproductive age; not just the elderly, infirm, or young like most infections. 'Those are your breadwinners, your mothers, your fabric of a viable society,' she says. 'Its social impact was overwhelming.' The fact it struck high-income countries as well as low and middle-income countries further galvanized support to stop it.

HIV is an extremely clever and manipulative virus that transmits via bodily fluids, including blood, breast milk, semen and vaginal secretions. Believed to have crossed over from monkeys, the virus uses a person's own cells to replicate and then, once virus levels are high enough, attacks their immune system, weakening their ability to fight infections and some cancers. It does this by depleting levels of certain white blood cells, called CD4 cells, and if left untreated, levels of these cells become so low that a person develops AIDS. While infections do predominantly spread sexually, it's important to note that a significant proportion of infections spread through contaminated needles – by drug users mostly, but occasionally poor healthcare practices – and from mother to child, bringing another face to the epidemic.

HIV spread like wildfire during the 1980s as experts struggled to understand its biology and find effective treatments. By 1990, an estimated 7.9 million people worldwide were living with HIV, according to data from the Joint United Nations Programme on HIV/AIDS (UNAIDS), and 290,000 died that year alone from AIDS-related conditions. That same year, an estimated 1.9 million adults and children were newly infected, with that number going up by hundreds of thousands each year. The peak came in 1996, when new infections reached a plateau of 2.9 million for four consecutive years. To highlight the true global nature of the epidemic, and the significant impact on industrialized countries, more than half a million people in the US had died from AIDS by that point.

New HIV infections only began to decline in 2000, at which point there were 24.9 million adults and children already living with HIV, an extraordinarily high number for a disease that had only surfaced twenty years earlier. But 1996, when new infections plateaued, also marked a significant year for controlling the epidemic. It was the year antiretroviral therapy (ART) as we know it today began. It involved using a combination of three drugs to suppress the ability of the HIV virus to replicate, bringing viral levels down to a minimum, while also reducing the chances of drug resistance developing. A common drug that had been used to treat infections before this point was azidothymidine (AZT), which did suppress the virus and reduce deaths, but brought with it some serious, life-threatening side effects, such as liver problems, muscle weakness and blood disorders, and doctors soon saw resistance to the drug emerge, as often happens when treatment relies on a single drug.

Today, over 30 drug options exist, of which physicians prescribe a combination therapy to suppress the HIV virus in those infected. Studies have shown repeatedly that treatment in this way makes the virus undetectable, enabling people to live a long and healthy life, despite their infection. The same studies suggest people are also highly unlikely to transmit the virus when levels are this low. Together, these are proving to be crucial elements in controlling the epidemic moving forward – along with the development of instant testing kits to more readily find those infected and get them on treatment.

'The first 35 years of the epidemic have been ones of containment,' says Bekker. 'The disease had no cure, could kill people, and so the virus itself needed to be controlled.' Now, the advent of effective drug regimens that make the virus undetectable has changed the course of the epidemic, moving it into an era 'where we're dealing with a long-term chronic disease', Bekker says, adding that now, 'global health teams need to find the 38 million people living with the virus, making sure they are on treatment in good health systems so they can access treatment throughout their lives and prevent any further transmission.' Those working to control HIV have now moved their focus to prevention strategies. These include messaging around safe sex and drug use, promoting procedures such as male circumcision – which has been shown to reduce the risk of infection in men by about 60 per cent – and more recently the use of certain antiretroviral drugs as a means of prevention in high-risk groups, such as gay men, called pre-exposure prophylaxis (PrEP). But each approach has its own flaws, mostly around compliance and access, with PrEP proving to be controversial politically and bringing fears of drug resistance emerging.

Bekker highlights an important point about the HIV epidemic today when she describes the shift to dealing with a 'long-term chronic disease'. Infection with HIV is lifelong. Treatment is effective, but it is not a cure and is unlikely to ever be. Most experts, including Bekker, now agree that a cure is off the cards and that, at best, we may achieve long-term remission, also known as a 'functional cure', where people can come off their drugs for a prolonged period of time or delay treatment. The rare cases of cures we have seen to date are unlikely to become the norm, both in terms of the science involved and the practicalities of performing the procedures required. For example, an American patient, Timothy Ray Brown, was able to stop HIV treatment after chemotherapy and bone marrow transplants to treat his cancer left him with cells resistant to HIV infection. But bone marrow transplants aren't going to become standard for people living with HIV as the risks involved in such a procedure outweigh taking a daily pill. A second case similar to Brown's cure was reported in 2019, however, giving some experts hope of understanding a form of cure that could be effective in certain HIV patients.

However, Bekker is more positive about the likelihood of an HIV vaccine, she says. It would be an understatement to say that developing a vaccine to prevent HIV would be a challenge, given the damage the virus causes to the human immune system. But there is a range of candidate vaccines out there and one has shown promise in clinical trials. A trial using a combination of two candidate HIV vaccines, ALVAC and AIDSVAX B/E was tested in Thailand, with the first priming an immune response and the second then boosting it. The trial found that this combo reduced the risk of contracting HIV within one year by 60 per cent, going down after three-and-a-half years to 31.2 per cent. This is the only vaccine to date to show some level of protection against HIV in a clinical trial, but it gives hope that a vaccine could be possible. 'The challenge now becomes around enhancing that and improving on that,' says Bekker. 'That is a slightly more attainable goal.' But she also points out that rolling out a vaccine that requires both a prime and a boost dose will bring its own challenges in terms of delivery through health systems, particularly in low- and middle-income countries. The Thai results led to a larger trial in South Africa with a vaccine adapted to that population, where a different subtype of the virus also occurs and rates of infection are significantly higher. The trial began in 2016 on a little over 5,400 HIV-negative volunteers at fourteen sites across South Africa but was halted in 2020 when preliminary analyses found it did not protect against infection.

But the search will continue. 'A vaccine gives us a chance to really control the virus around the world,' says Bekker.

Where do things stand today with infection rates? Some countries are close to having no new infections, with fewer than 100 reported each year. These are largely in Western Europe and North America. Sub-Saharan Africa, which bears the brunt of infections, is finally seeing a decline in new infections and AIDS-related deaths, but still has a long way to go. Alongside this positivity is some negativity: in Eastern Europe, Central Asia, the Middle East, North Africa and Latin America, infections are going up.

The world had a target, set by UNAIDS, dubbed 90-90-90. This was the ambitious goal of 90 per cent of all people living with HIV knowing their HIV status, 90 per cent of all people with diagnosed HIV infection receiving sustained antiretroviral therapy, and 90 per cent of all people on antiretroviral therapy reaching viral suppression, all by 2020. A recent report by UNAIDS, published in summer 2019, found that progress was not where it should be and that goals were unlikely to be met. South Africa, for example, which has the most people living with HIV of any country, has reduced new infections by 40 per cent since 2010. But there were still an estimated 7.7 million people living with HIV there in 2018. Meanwhile, Eastern Europe and Central Asia saw the greatest percentage spike in new infections, as well as AIDS-related deaths, of any region. It should be noted that the actual numbers in Eastern Europe and Central Asia are markedly lower when compared to Africa, with 150,000 people newly infected in 2018 compared with 800,000 in East and Southern Africa, but the worry is that numbers are going up, aided by lack of access to prevention measures and a notable epidemic among injecting drug users. Unless we turn what's happening in Central Asia around, it's an even longer timeframe until we control the epidemic because we haven't even begun to stop infections there, explains Bekker.

Also in the picture at the moment is another even more ambitious goal: to end the AIDS epidemic by 2030. Bekker believes this target is disingenuous, creating the wrong message around control efforts. 'We are nowhere near the end of AIDS,' she says, believing it's not helpful to talk about the end of AIDS until we have a prophylactic vaccine, and when more than 38 million people currently need to be on antiretrovirals for the rest of their lives. That's right, tens of millions of people currently need to access drugs to keep their HIV levels suppressed for the rest of their lives, while millions continue to get newly infected each year. Of those new infections, around 500 a day are among babies born with the virus, who will need to be on treatment for their entire lives. The route to truly controlling HIV is unbelievably long, and the route to ending it, seemingly infinite, involving multiple lifetimes. 'We have to bat down for the long haul,' says Bekker. 'Unless we get a prophylactic vaccine or a cure within this next generation or two, this thing is with us for quite a few more generations.'

EPILOGUE

Humanity has always been battling infection – and it always will. Diseases that once wiped out entire communities, like the plague or syphilis, may have calmed down today, but they are still infecting people somewhere in the world and in some cases are seeing resurgences or local outbreaks. We will never truly win the battle against all infectious diseases, but instead, as a species, we adapt our fight as the scenarios change – because it is the way in which we experience infections that has changed. Poor hygiene and living conditions once enabled simple infections to spread easily, but improved environments and the advent of antibiotics helped curb these significantly. Today, the challenges are different, but many: the global population edges towards 8 billion and claims new territory to give all those people somewhere to live, in turn increasing their proximity to wild animals; there is constant population movement; we see overuse and poor management of drugs; and the ability for everyone in the world to communicate with anyone else in the world – also known as the internet – is enabling the spread of misinformation.

These factors determine the modern-day battle against disease and two words bring them all together: globalization and technology.

As the European Centre for Disease Prevention and Control's expert Dr Jan Semenza has said, globalization is here to stay and there's no point trying to stop it. The same can be said of technology. Efforts need to continue to monitor and control messaging around public health – in all parts of the world. All we can do is try to keep up and solve the problems that arise along this trajectory, like fatal outbreaks.

In a way, human nature is to blame for the longevity and persistence of the infections we face today. It is human nature to explore the world, to push limits, move into new areas and be sociable. This takes us either to remote locations, where health services struggle to reach us, to dense cities where contagion of some kind is inevitable, or closer to nature itself in the form of wild animals that then bring us new infections. Wherever we go there is risk. It is also human nature to take shortcuts or seek a quick fix to restore our health or provide food as soon as possible, in this case leading to the overuse or misuse of our beloved antibiotics on both humans and farm animals. It's in our psychology to constantly question the world around us and often lead with our emotions and sentiment over rational thought, giving considerable power to seeds of doubt once sown. Pockets of people have long doubted vaccines, but today their voices have the ability to spread suspicion globally with ease, aided by a climate in which people are also questioning the authority of experts and the media, growing those pockets into full blown communities. In some circumstances, one individual's voice has been as powerful as decades of scientific evidence, resulting in worldwide outbreaks of preventable diseases.

In the face of all of this, global health teams persist, dedicating their time, energy and resources to keep levels of each disease at whatever minimum they can achieve. Surveillance, containment, prevention, new drugs and vaccines, rapid diagnostics; whatever it takes to bring the numbers down and under control. Using such measures, smallpox has been eradicated; Guinea worm and polio are close, but likely to stay that way for some time. For other diseases, the goal is simply to stop them being too big a public health problem. Eradication or elimination is the ideal, though more often a slow decline over time and the prevention of large-scale outbreaks is the reality. Keeping those affected informed, engaged and in receipt of the right information is a new element to this reality, but all combined, this is how the battle has developed in the twenty-first century. By the twenty-second, who knows what we'll be facing and what tools we'll have to fight back with, but it's fair to say there will still be infections to beat and it will likely be human behaviour that decides which ones.

ACKNOWLEDGEMENTS

I want to thank my family, without whom I would not have had the time, space or determination to write this book. Thank you to my Amma for her endless love and support (and babysitting) as I wrote; to my Appa, whose passion for public health drives my own; my husband for being my rock and sounding board through the entire creative process; and my son, whose arrival pushed me to write and (hopefully) inspire him in the future.

I want to also thank the team at Icon Books, notably Brian Clegg, Duncan Heath and Robert Sharman, who not only gave me the idea for this book, but also provided endless flexibility and support as I wrote it following the birth of my son.

I care deeply about the field of global health and communicating the various health challenges that face communities, rich and poor, worldwide. Thank you to everyone who has played a part in my training and journalism career to date for enabling me to do this.

FURTHER READING

Books

Heymann, David (ed.), *Control of Communicable Diseases Manual*, 20th edition (Washington DC: APHA Press, 2015)

Hopkins, Donald R., *The Greatest Killer: Smallpox in History* (Chicago: University of Chicago Press, 2002)

Kucharski, Adam, *The Rules of Contagion: Why Things Spread – and Why They Stop* (London: Profile, 2020)

Larson, Heidi J., *Stuck: How Vaccine Rumors Start – and Why They Don't Go Away* (New York: Oxford University Press, 2020)

Piot, Peter, *No Time to Lose: A Life in Pursuit of Deadly Viruses* (New York: Norton, 2012)

Websites

World Health Organization
www.who.int

Centers for Disease Control and Prevention (US)
www.cdc.gov

European Centre for Disease Prevention and Control
www.ecdc.europa.eu

The Vaccine Confidence Project
www.vaccineconfidence.org

Wellcome Global Monitor
wellcome.ac.uk/reports/wellcome-global-monitor/2018

Global Polio Eradication Initiative
polioeradication.org

The Carter Centre, Guinea Worm Eradication Program
www.cartercenter.org/health/guinea_worm

INDEX